Energy ... Beyond Oil

Energy ... Beyond Oil

Fraser Armstrong and Katherine Blundell

OXFORD
UNIVERSITY PRESS

OXFORD

UNIVERSITY PRESS

Great Clarendon Street, Oxford OX2 6DP

Oxford University Press is a department of the University of Oxford.
It furthers the University's objective of excellence in research, scholarship,
and education by publishing worldwide in

Oxford New York

Auckland Cape Town Dar es Salaam Hong Kong Karachi
Kuala Lumpur Madrid Melbourne Mexico City Nairobi
New Delhi Shanghai Taipei Toronto

With offices in

Argentina Austria Brazil Chile Czech Republic France Greece
Guatemala Hungary Italy Japan Poland Portugal Singapore
South Korea Switzerland Thailand Turkey Ukraine Vietnam

Oxford is a registered trade mark of Oxford University Press
in the UK and in certain other countries

Published in the United States
by Oxford University Press Inc., New York

British Library Cataloguing in Publication Data

Data available

Library of Congress Cataloging in Publication Data

Data available

Typeset by Newgen Imaging Systems (P) Ltd., Chennai, India
Printed in Great Britain
on acid-free paper by
Biddles Ltd., King's Lynn

ISBN 978–0–19–920996–5

10 9 8 7 6 5 4 3 2 1

Contents

Acknowledgments

Warm thanks are due to Dr Tony Boyce, Bursar of St John's College, Oxford, for his support for the initial one-day workshop "Energy . . . beyond Oil", held at St John's College in 2005, from which this book emerged. It is also a pleasure to thank the participants of the Discussion Panel at that workshop for their stimulating and insightful comments. The Panel Members were: Brenda Boardman, Howard Dalton, John Gummer, John Hollis, Robert Mabro, Michael Scholar, Simon Weeks and David Vincent. We are grateful to Katrien Steenbrugge and Nick Jelley for helpful comments on the manuscript.

Fraser A. Armstrong, Katherine M. Blundell, Oxford, July 2007.

All Author's royalties for this book are being donated directly to Oxfam, an organisation that works with others to overcome poverty and suffering around the world.

1. *Energy ... beyond oil: a global perspective*

Fraser Armstrong, Katherine Blundell, and Ian Fells

The problems to be solved

Coal and oil, which are the buried products of several hundred million years of solar energy, photosynthesis, and geological pressure, have fuelled our industries and transport systems since the Industrial Revolution, a period of only 200 years. Although opinions differ as to when the peak in oil production will occur (perhaps in 2010, perhaps in 2030), it is hard to avoid the conclusion that oil is being consumed about one million times faster than it was made and, further than this, the twenty-first century, will be the century when societies have to learn to live without gas and oil (coal will outlast oil and gas by a few hundred years). But, there is an entirely separate motivation for living without fossil fuel: obtaining energy from oil, coal, and gas will continue to put carbon dioxide (CO_2) into the atmosphere at levels which it is widely acknowledged are elevating the average temperature on the planet. Carbon dioxide is a good heat absorber and acts like a blanket: this is because CO_2 molecules resonate strongly with infrared radiation causing it to be trapped as heat instead of all being transmitted into space. Global warming is already causing the polar ice caps to melt and it is inevitable that there will be higher sea levels resulting in less land for an increasing population, along with changes in climate. These changes are not easy to predict and may be difficult to reverse. Either of these two motivations, be it the depletion of oil reserves or the need to arrest global warming caused by the combustion of fossil fuels, mandates new thinking from all those with concern for the future.

How will future generations view our policies and our decision making today? Unless we change course now, these people will be left in a world where energy is a scarce resource and the mobility we have taken for granted in the late twentieth and early twenty-first century will be long gone. Our generation—rightly—would be blamed for knowingly squandering the planet's resources. We would have burnt all the fossil fuels formed in the history of the world in an orgy of combustion lasting only a few hundred years. There would be no precedent for this; hitherto, no generation would have caused future generations to look back in such reproach.

The laws of thermodynamics that govern the supply and conservation of energy are well understood. Even carbon sequestration, which might stem the increase of CO_2 in the atmosphere, requires energy. Saving energy, though laudable and imperative, is insufficient to solve the problems that lie ahead: oil will run out and then gas and then coal, stunting the growth of the developing world. Moreover, there is no moral high ground for the developed world to stand upon and require of the developing world that they adopt principles far greener than we have held ourselves.

How can we supply energy for the inhabitants of Earth, sufficient for them to enjoy reasonable living standards, without causing serious, perhaps irreversible, damage to the environment? To achieve a 1% growth in the economy of a developing country requires a 1.5% increase in energy supply. China, whose Gross Domestic Product is growing at 7–10% per year, commissions a new power station every week!

What are the actual needs?

The total global annual energy consumption at present is 10,537 million tonnes of oil equivalent[1] of which the EU consumes 1,715 million tonnes and the USA 2,336 million tonnes.[2] Based on current projections, the global annual energy consumption rate will double the current figure by 2050 and triple or perhaps even quadruple by the end of the century. About 85% of the total global energy consumed at present comes from burning fossil fuels, with the proportion approaching 90% for developed countries. The remaining sources of energy are

[1] 1 tonne of crude oil = 42 Giga Joules = 7.3 barrels of oil. 1 Terawatt hour (TWh) $= 3.6 \times 10^{15}$
    Joules ; 1 million tonnes of oil produces 4.5 Terawatt hours of electricity (based on 40% efficiency).
[2] Data from BP statistical review of World Energy.

Table 1.1 Current world and UK usage of different energy sources (GToe and MToe are Giga and Mega tonnes of oil equivalent, respectively).

Energy source	World usage (GToe)	UK usage (MToe)
Coal	2.2	40.3
Oil	3.5	76.1
Gas	2.2	85.9
Total fossil	7.9	202.3
Nuclear	0.6	21.3
Renewables, commercial	0.6	1.5
Biomass, non-commercial	1–2	Very small

Source: 'Ingenia' R. Acad. Eng. Issue 17 (2003).

Table 1.2 World electricity generation from different sources.

Source	% of world production
Coal	38.1
Gas	17.1
Oil	8.5
Nuclear	17.2
Hydro	17.5
Other (solar PV, wind)	1.6

Source: 'Ingenia' R. Acad. Eng. Issue 17 (2003).

hydroelectric, nuclear, biomass and other renewables—such as solar, wind, tide, and wave. Table 1.1. shows the current world (and UK) usage of different energy sources while Table 1.2 shows what the main sources of electricity are.

What are the true costs of the different energy solutions in terms of human fatalities?

Table 1.3 lists the dangers of different human activities, quantified as loss of life expectancy (LLE) in days. The danger from nuclear energy is significantly less than is often portrayed. Figure 1.1 indicates the human fatalities resulting from different energy sources.

When the cost of energy is considered in deaths per TW-year, it is clear that nuclear fission is considerably less dangerous than gas, coal or (especially) hydroelectric energy.

Table 1.3 Dangers of different human activities measured in terms of the loss of life expectancy in days for a 40-year old person.

LOSS OF LIFE EXPECTANCY

Activity of risk	Days LLE
Being male rather than female	2,800
Heart disease	2,100
Being unmarried (or divorced)	2,000
Cigarettes (1 pack/day)	1,600
Coal mining	1,100
30 lbs overweight	900
Alcohol	130
Small cars v. standard size	50
All electric power in US nuclear (UCS)	1.5
Airline crashes	1
All electric power in US nuclear (NRC)	0.03

Source: Professor Bernard Cohen, University of Pittsburgh, Dept. of Physics. (UCS denotes estimate made by the Union of Concerned Scientists while NRC denotes estimate by US Nuclear Regulatory Commission.)

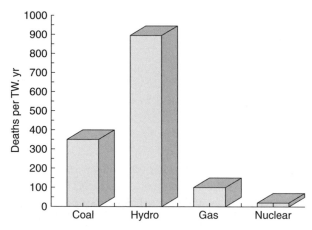

Figure 1.1   Deaths per TW.yr (i.e. normalized with respect to the amount of energy generated per year) from different energy supplies. Source: World Nuclear Association.

Fig 1.2 illustrates how nuclear energy puts significantly fewer kilogrammes of $CO_2 (\sim 1 \%)$ into the atmosphere, per kWh of energy, than either coal- or gas-fired power stations.

Fig 1.3 illustrates how much natural resource is available, when translated into units of Gigatonne of oil equivalent (GTOE) and how use of fast reactor technology multiplies world energy resources by ten. Using fusion technology could be even more efficient would give an even larger resource.

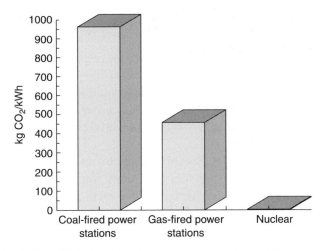

Figure 1.2 Production of CO_2 by different sources of energy for electricity generation. Source: World Nuclear Association.

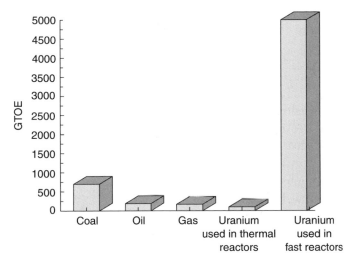

Figure 1.3 Availability of natural resources worldwide, measured in units of Gigatonne of oil equivalent (GTOE). Source: UKAEA.

Energy from the Sun

The Sun lies behind many sources of energy that are available to us. Energy from wind ultimately depends on the heating of the atmosphere by solar radiation, and a combination of gravitational and thermal effects within the atmosphere and ocean cause winds. Harnessing this kinetic energy of the air is already being realized as an energy source, using wind turbines.

The gravitational fields of the Moon and Sun are responsible for causing twice-daily tides in the oceans on Earth. In fact, if the surface of the planet were entirely covered by oceans the amplitude of the tides would only be about 0.4 metres, but the presence and structure of land masses causes tidal amplitudes of over 10 metres. Harnessing the kinetic energy of powerful tidal currents has begun in certain suitable places.

Most obviously, the Sun provides solar energy to our planet on an annual basis at a rate of $\sim 100,000$ TW. Therefore the energy from one hour of sunlight is equivalent to all the energy mankind currently uses in a year.

The nature of the solutions

It is not easy to match oil as an energy source: oil itself is cheap, energy-dense, convenient and easy to transport. Getting from renewable primary energy to energy-dense fuels is a particular challenge. It is a challenge to keep aviation and other forms of transport going, given that the production of ethanol from biomass should not be at the price of taking up huge stretches of land with a vast monoculture. The hydrogen economy has been mentioned as a future saviour, but hydrogen is not a primary energy source but rather an energy store. Water is the chemical product of the reaction of hydrogen and oxygen, which of course releases a lot of energy just as does the burning of fossil fuels. But unlike fossil fuels, inter-conversion between water and hydrogen is easily reversible (although energy intensive) so that primary energy sources such as sunlight, the intense heat provided by a high temperature fusion reactor, or indeed electricity generated by any means, can provide sufficient energy to split water into hydrogen and oxygen. Hence the definition of hydrogen as an energy store. We thus obtain a fuel that can be transported. Chemists are responding to these challenges by devising advanced materials that store hydrogen and even by reacting hydrogen with CO_2 to get back to hydrocarbons, the energy-dense fuel of choice (and now renewable of course!).

In this book *Energy ... Beyond Oil*, we consider energy solutions for which the science or technology is proven but still developing: geothermal energy, energy harnessed from the waves and tides (which of course have the advantage of being predictable) as well as energy from wind (we learn that deaths to birds from wind turbines are nothing compared to bird deaths from cats). The cases are presented for the two different types of 'nuclear' energy: fission and fusion. Unfortunately the word 'nuclear' often causes a knee-jerk reaction among those for whom the terminology simply brings to mind the image of a mushroom cloud. We remind the reader that it is easy to forget that solar energy is nuclear (fusion) just as geothermal energy actually arises partly from natural nuclear processes (the radioactive decay of uranium) that take place in the Earth's core. The case is made that fission, far from being the emblem of a future catastrophe, could instead save the planet. With improvements in reactor design and fuel technology, we have the option of an energy supply that is relatively clean compared to fossil fuel. Fusion is much cleaner still. Fusion as a physical mechanism is well understood: stars have been fuelled by fusion since shortly after the beginning of time. Fusion as a technology requires the dedicated time and talents of engineers and physicists who can mimic Nature and realize the potential of this clean, green, effectively limitless, energy supply.

Solar energy is of course harnessing fusion from a distance of 93 million miles. In fact the illuminated Earth receives on average about one kilowatt of power per square metre, easily enough in principle to solve all our energy needs. Photosynthesis—the process by which plants capture solar energy using light sensitive pigments and use this energy to make organic matter from CO_2 and water—is indeed the start of the food chain. Yet only a small fraction (less than 1%) of the Sun's energy is actually trapped and exploited in this way. Much is being done to understand and exploit photosynthesis (such as the specialized production of crops for energy) and to develop 'artificial photosynthesis' in which light is converted directly into electricity using solar photovoltaic panels. More straightforward, and widely used, is the direct exploitation of solar hot water panels for domestic heating. In the sunniest of locations it is also feasible to concentrate solar radiation using mirrors (concentrating solar power technology) to provide enough heat to drive turbines for electricity production.

The energy alternatives to oil should not be regarded as alternatives to one another, rather it is imperative to consider 'both . . . and' rather than 'either . . . or'.

Regardless of how successful we become in energy production, minimizing the various costs and risks is a strong motivation for energy efficiency. All the different energy sources have their advantages and strengths in certain situations; for example, tapping into a particular local energy supply has many advantages for places distant from a large metropolitan area. It is important to offer locally optimal solutions in areas of low population density. Apart from wisdom in optimal resource use, the implementation of a number of energy solutions across any given nation limits the possibility of single-point failures in terms of vulnerability to terrorist attack or distribution breakdown. Security of supply and protection of the environment must remain paramount.

The way forward

The question of *Energy . . . Beyond Oil* brings together scientists of all disciplines: chemistry, physics, biology, materials, engineering, as well as politicians and industrialists. Not to think about this question is to be in denial about the reality of the severe and looming problems that lie ahead. The costs of alternative energy solutions are dropping, but even so we need increased investment from both private and public sectors, along with government-led incentives, to see us through to the time when fossil fuels are no longer the automatic choice. There are important opportunities for young scientists and engineers—professionals who love challenges and problem solving. *Energy ... Beyond Oil* lays out the greatest challenge for this century.

The Authors

Professor Fraser Armstrong is Professor of Chemistry in the Department of Chemistry, Oxford and a Fellow of St John's College. His interests are in inorganic chemistry, biological chemistry, bioenergetics, and in the mechanisms and exploitation of enzymes related to energy production. He has received a number of awards including the European Award for Biological Inorganic Chemistry, the Carbon Trust Innovation Award, the Max Planck Award for Frontiers in Biological Chemistry, and the Royal Society of Chemistry Award for Interdisciplinary Chemistry. He travels widely giving invited lectures on topics including catalysis, bioenergetics, and renewable energy.

Dr Katherine Blundell is a Royal Society University Research Fellow and Reader in Physics at Oxford University, and a Science Research Fellow at St John's College, Oxford. Her interests include extreme energy phenomena in the Universe, for example, around black holes. She is frequently invited to speak at conferences and different institutes around the world and has published extensively on astrophysical jets, relativistic plasmas, and distant galaxies. She has co-authored the book *Concepts in Thermal Physics* (OUP, 2006) and was recently awarded a Leverhulme Prize in Astronomy & Astrophysics.

Professor Ian Fells CBE FREng FRSE is Emeritus Professor of Energy Conservation at Newcastle University. He has published some 290 papers on energy topics as varied as fuel cells, rocket combustion, chemical physics of combustion, nuclear power, and energy economics. He was science advisor to the World Energy Council and to select committees of both Houses of Parliament. He has made over 500 TV and radio programmes and was awarded the Faraday Medal and Prize by the Royal Society for his work on explaining science to the layman. For some publications see www.fellsassociates.com.

2. *Arresting carbon dioxide emissions: why and how?*

David Vincent

Introduction

This chapter sets the scene for future chapters covering a range of low carbon technologies from renewables through to nuclear. It reviews how the evidence base for climate change is building up, what the impacts of climate change might be, and how we are beginning to explore the policies and measures which will be needed to make the transition to a low carbon economy.

The year 2005 will go down in history as the beginnings of a broad, politically-backed consensus that man's activity is influencing our climate. In February 2005, the Kyoto Protocol came into force—binding over 170 countries in action to reduce carbon dioxide emissions, accepted by most informed commentators to be the principal cause of anthropogenically forced climate change. In the same year, the G8 group of countries at Gleneagles, Scotland, considered climate change as a key agenda item. Significantly, it set up a forum for discussion with other countries and the emerging economies. The forum, known as the 'Dialogue on Climate Change, Clean Energy and Sustainable Development' met for the first time in November 2005.

However, the value of the Kyoto protocol is not universally acknowledged. Some argue that although the science underpinning the existence of climate change and the link with carbon dioxide emissions has become unequivocal, the Kyoto protocol is not appropriate for them. A group of these countries, including the US, China, and India (huge emitters of carbon dioxide in their own right) has agreed the need to tackle climate change. Their approach is to promote clean technology development initiatives; though how exactly that partnership will evolve and deliver new low carbon technologies is not, at the time of writing, clear. Nevertheless, whether via the formalized Kyoto Protocol with carbon dioxide emission reduction targets or via other initiatives, a start has been made on the long, uncertain road to a low carbon world. Slowly, but surely, global action on climate change is gathering momentum.

Principles

The greenhouse effect

The term 'greenhouse effect' was first coined by the French mathematician Jean Baptiste Joseph Fourier in 1827. It enables and sustains a broad balance between solar radiation received and Earth's radiation emitted or reflected. Without that broad balance, temperatures would be about 33°C cooler and life, as we know it, would not exist. Fig. 2.1 shows how that broad balance arises and is maintained.

The sun's radiation penetrates our atmosphere. Visible and ultraviolet radiation from the high-energy end of the spectrum penetrates most efficiently and directly warms the earth. Re-radiation from the Earth's surface is mainly from the infra-red, lower-energy end of the spectrum radiation, but a significant portion of this radiation is absorbed by certain molecules in the atmosphere that convert it into heat, thereby keeping the earth at temperatures that can support life. Significantly, carbon dioxide is a particularly good absorber of infra-red radiation.

In 1859, the British scientist John Tyndall began studying the radiative properties of various gases. He suggested that it was variations in carbon dioxide levels that brought about the various 'ice ages' in the Earth's geological history. The cyclical occurrence of ice ages has been established from analysis of Antarctic ice core samples, with data going back about 800,000 years: temperature variations of between 5–10°C are estimated to have occurred. The lower temperatures correspond to ice ages, and the higher temperatures to warm periods such as the one we are experiencing at present. Tyndall attributed these

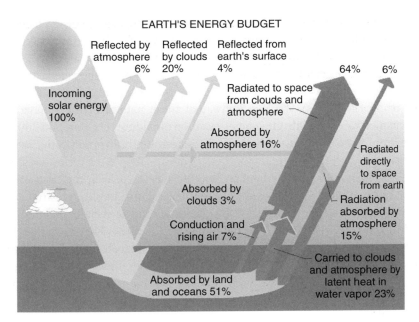

Figure 2.1   The Earth's energy budget. Source: NASA.

temperature changes to variations in atmospheric carbon dioxide concentrations. Changes in our atmosphere, principally carbon dioxide concentrations, water vapour, and particulates (for example, from volcanic activity), can change the efficacy of the absorption process and hence the temperature of the earth's surface.

In 1896, the Swedish chemist Svante Arrhenius made the first attempt to estimate the effect of carbon dioxide on global average temperatures. Using a simple physical model, he estimated that if atmospheric carbon dioxide concentrations were to be doubled the average global temperature would rise, due to the greenhouse effect, by an estimated 5–6°C—an estimate which turned out to be not very different from the most recent model of the Inter-governmental Panel on Climate Change (IPCC) in its Third Assessment Report (IPCC, 2001).

Scientific evidence of the existence of climate change

The IPCC was established in 1988 by the World Meteorological Office and the UN Environment Programme to assess scientific, technical, and socio-economic information relevant for the understanding of climate change. It said in its Third Assessment Report, published in 2001, that: 'an increasing body of observations gives a collective picture of a warming world and other changes in the climate system'. The IPCC's Fourth Assessment Report is due to be published in 2007.

There is an almost universal scientific consensus that the climate change we are seeing now, and which is beginning to change the natural and economic environments in which we all live, is largely due to human activity. Natural climate change has ebbed and flowed over hundreds of thousands of years with warmer periods interspersed with cooler periods every 12,000 to 20,000 years or so (Fig. 2.2). We also know that these natural changes have been accompanied by changes in atmospheric carbon dioxide levels—around 200 parts per million (ppm) during cooler periods and around 280 ppm during warmer periods.

In less than 200 years, human activity has increased the atmospheric concentration of greenhouse gases by some 50% relative to pre-industrial levels. At a little over 380 ppm, today's atmospheric carbon dioxide concentration is higher than at any time in at least the past 420,000 years. There is no previous human experience of the Earth's atmosphere at current levels of greenhouse gases to assist us in predicting the consequences. It is likely, though, that the natural

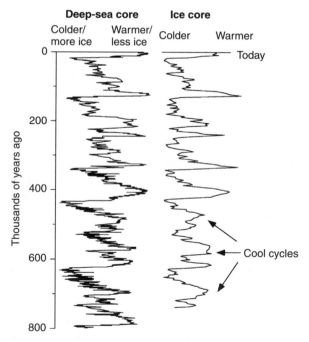

Figure 2.2   Glacial cycles of the past 800,000 years. The history of deep-ocean temperatures and global ice volume is inferred from a high resolution record of oxygen-isotope ratios measured in bottom-dwelling foraminifera shells preserved as microfossils in Atlantic Ocean sediments. Air temperatures over Antarctica are inferred from the ratio of deuterium and hydrogen in the ice (McManus, 2004).

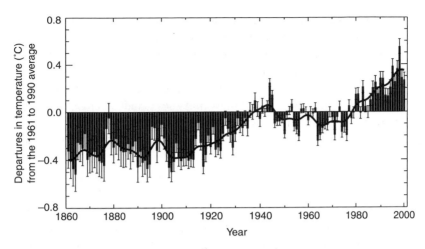

Figure 2.3 Variations in the earth's surface temperature over the past 140 years since instrumented records began. (IPCC, 2001).

oscillating pattern of ice ages and warm periods is now being disturbed in ways and with impacts we currently do not understand.

This massive and rapid rise in carbon dioxide levels is very largely attributed to the burning of fossil fuels to generate energy and to provide transport fuels. Not only is this unprecedented in absolute terms but also the *rate* of change is faster than has ever been observed before. This rapid change is occurring during a period of human development characterized by massive populations and massive demands on finite resources, including energy.

Fig. 2.3 shows direct temperature records back to the middle of the nineteenth century which are considered reliable enough to establish the fact that recent temperatures are warmer than any time since direct measurements began. All of the 10 warmest years have occurred since 1990, including each year since 1995. On 22 October 2005, the UK experienced the highest October temperatures since records began nearly two centuries ago.

Fig. 2.4 shows the change in atmospheric concentrations and radiative forcing functions for three of the main greenhouse gases: carbon dioxide, methane, and nitrous oxide.

In contrast to natural climate change, anthropogenically forced climate change is a new phenomenon caused, predominantly, by our use of fossil fuels to power developed (and, increasingly, developing) economies. Since the start of the industrial revolution some years ago, we have been increasing the concentration of carbon dioxide and other greenhouse gases in the atmosphere, thickening the greenhouse blanket and beginning the inexorable rise in global temperatures.

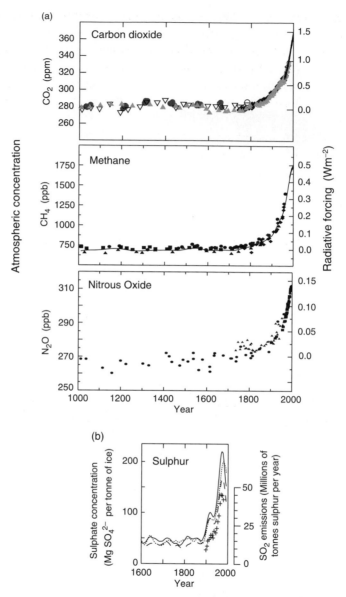

Figure 2.4 Global atmospheric concentrations of three of the principal greenhouse gases. (IPCC, 2001).

It took many millions of years to trap carbon chemically in geological strata. Modern man is releasing, in the 'twinkling of an eye' relatively speaking, huge quantities of carbon dioxide—with effects we are now beginning to see. Carbon dioxide levels have increased by 31% since 1750. The present atmospheric carbon dioxide concentration (over 380 ppm—and rising) has not been exceeded during the past 420,000 years and is likely not to have been exceeded during the past 20 million years. The current *rate* of increase is unprecedented during at least the past 20,000 years. Bearing in mind the historically observed atmospheric carbon dioxide concentration range (around 200–280 ppm), we are, quite simply, in uncharted territory so far as greenhouse gas concentrations and their impacts are concerned.

With such a complex subject as climate change, there are areas where our understanding needs to improve—for example, the 'radiative forcing' impacts of aerosols and mineral dusts, and the ability of oceans to absorb carbon dioxide. However, we know enough to be sure that global warming over the past 100 years is very unlikely to be due to the variability of natural phenomena alone. In fact, the best agreement between model simulations and observations is found when anthropogenic and natural forcing factors are combined, as shown in Fig. 2.5.

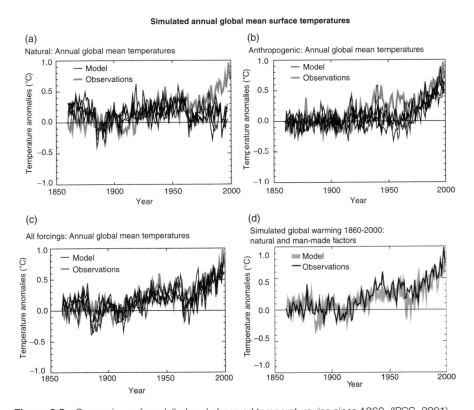

Figure 2.5 Comparison of modelled and observed temperature rise since 1860. (IPCC, 2001).

A. Circa 1900   B. Recent
Photo Source: Munich Society for Environmental Research

Figure 2.6   Pasterze glacier around 1900 and in 2000—Kartnen, Austria. (Source: Munich Society for Environmental Research.)

Figure 2.7   Accelerated melting of parts of the Antarctic ice cap. (Source: BBC News website, March 2002.)

It is very likely that the twentieth-century warming has also contributed to a rise in sea levels, through thermal expansion of sea water and widespread loss of land ice. Fig. 2.6 shows an example of glacial retreat and Antarctica provides another example of climate change (Fig 2.7).

In summer 2003, as shown in Fig. 2.8, a heat anomaly across Europe was responsible for 26,000 deaths and cost around €13bn. This was closely repeated in 2005.

An idea of year-on-year variations is given in Fig. 2.9 which shows summer-time temperature data from 1864 to 2003.

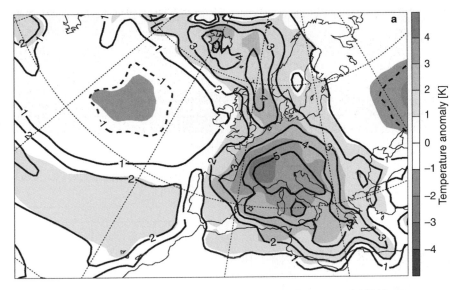

Figure 2.8 Localized temperature anomaly—summer 2003. (Schaer *et al.* 2004.)

Future projections on climate change

Looking ahead, what are the likely trends in global temperatures and climate change? Are we going to experience far more severe weather events on a global scale? What will their impacts be?

Emissions from fossil fuel burning are virtually certain to drive the upward trend in atmospheric carbon dioxide levels. The IPCC Third Assessment Report projects carbon dioxide levels in the range 490–1260 ppm by the end of this century. Correspondingly, the globally averaged surface temperature is projected to increase by 1.4 to 5.8°C over the period 1990–2100 as shown in Fig. 2.10.

Of equal importance are projections of changes in global precipitation. On a global scale, water vapour, evaporation, and precipitation are projected to increase. However, at regional levels the various models show both increases and decreases in precipitation. For high-latitude regions precipitation will increase in both summer and winter. Increases in winter precipitation are predicted for northern mid-latitudes, tropical Africa and Antarctica, and increases in summer precipitation are predicted for Southern and Eastern Asia. On the other hand, Australia, Central America, and Southern Africa are projected to show consistent decreases in winter rainfall.

The science of modelling climate change is developing all the time; reliability is improving and predictions of likely impacts are becoming more certain. The UK's

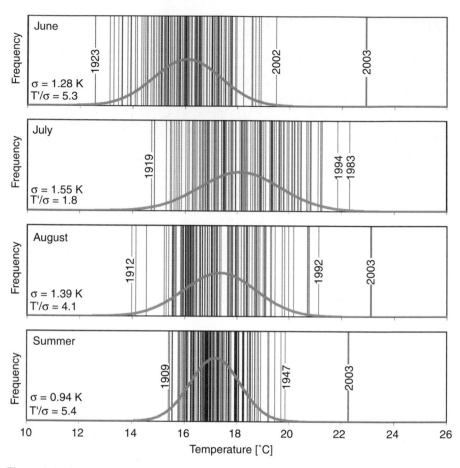

Figure 2.9 Distribution of Swiss monthly temperatures for June, July, and August, and seasonal summer temperatures for 1864–2003 (Schaer *et al.* 2004). The fitted Gaussian distribution is shown as a solid curve. The values in the lower left of each panel list the standard deviation (σ) and the 2003 anomaly normalized by the 1864–2000 standard deviation (T^1/σ). For summer 2003, the standard deviation was 5.4—which means that it is very unlikely that the temperatures recorded in summer 2003 were within the normal pattern of summertime temperatures for the past 150 years. The European summer temperature trend since 1900 has been rising.

Hadley Centre for Climate Change Prediction and Research at Exeter is widely acknowledged to be among the world's leading centres of excellence. Conclusions of a symposium in February 2005 entitled 'Avoiding Dangerous climate change—A Scientific Symposium on Stabilisation of Greenhouse Gases', were: (i) the pace of climate change exceeded that thought to be the case a few years ago; and (ii) the possibility of more extreme climate change was greater than expected from models developed a few years ago (Schellnuber, 2005).

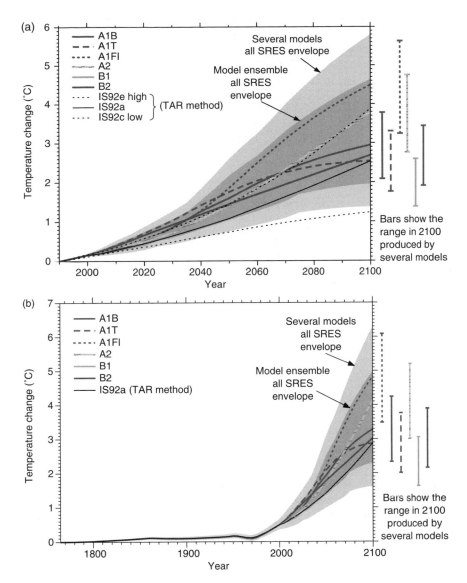

Figure 2.10 Simple model results: (a) global mean temperature projections for the six illustrative Special Report on Emissions Scenarios (SRES) scenarios using a simple climate model tuned to a number of complex models with a range of climate sensitivities. (IPPC, 2001) (b) As for (a) but for a longer timeline.

The effects of anthropogenically forced climate change will persist for many centuries. Emissions of long-lived greenhouse gases (not only carbon dioxide but nitrous oxide, perfluorocarbons, and sulphur hexafluoride) have a lasting effect on atmospheric composition, radiative forcing, and climate. For example,

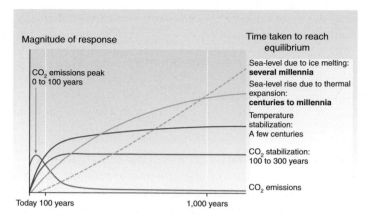

Figure 2.11   Accumulating impacts of climate change over the long term. (Adapted from IPCC, 2001).

several centuries after carbon dioxide emissions occur, about a quarter of the increase in carbon dioxide caused by these emissions is still present in the atmosphere. So even if emissions due to human activity were very significantly reduced over the next 100 years, peaking, say, 30 years from now, carbon dioxide would continue to accumulate over the next 100 to 300 years, and would remain roughly stable for the remainder of the millennium. The long lifetime of carbon dioxide in the atmosphere is, therefore, a significant factor governing climate change impacts. Fig. 2.11 shows how key aspects of climate change will continue to accumulate long after global emissions are reduced to low levels. Temperatures will continue to rise slowly for a few centuries as the oceans continue to warm. Sea level rises will continue for hundreds to thousands of years, due to the continuing impact on ice sheets and the thermal expansion of the oceans.

What atmospheric level of carbon dioxide is 'safe'?

There is no straightforward answer to this question—and, indeed, it begs the further question 'safe for whom—or for what?'. Much depends on our state of knowledge of the possible range of climate change impacts and that, in turn, depends on observed data and modelling projections. In his address to the Royal Institute of International Affairs in October 1998, Professor Michael Grubb, one of the UK's foremost expert commentators on climate change, said:

'I believe that policy at present should be guided by the objective of ensuring that we can, if necessary, stabilize atmospheric carbon dioxide concentrations in the range 450–550 ppm carbon dioxide, which equates to a range of about 500–650 ppm of all greenhouse gases, spanning a range broadly around a doubling of pre-industrial greenhouse gas concentrations.'

In 2002, the Government's Chief Scientific Adviser, Sir David King, said:

'If we could stabilise carbon dioxide levels to, say, around 550 ppm (which is around twice pre-industrialisation levels), current models suggest that there would be a significant mitigation of the effects of climate change.'

At the Hadley Centre scientific symposium of 2005, referred to above, there was much cause for concern. The international scientific community on climate change reported a succession of observed and predicted changes to our climate. The full report of the symposium was published in January 2006. Observed changes reported include:

(i)　0.6°C rise in annual average global temperatures;

(ii)　1.8°C rise in Arctic temperature;

(iii)　90% of Earth's glaciers retreating since 1850;

(iv)　increased freshwater flux from Arctic rivers appears to be already at 20% of the levels which are estimated would cause shutdown of thermo-haline circulation (THC);

(v)　Arctic sea ice reduced by 15–20%.

Predicted changes included:

(vi)　at around 1.5°C rise above pre-industrial temperatures, we could see an onset of complete melting of Greenland ice—causing, when complete, about 7 m of additional sea level rise;

(vii)　at around 2–3°C rise above pre-industrial temperatures (equivalent to about a doubling of atmospheric carbon dioxide concentrations) we could see the conversion of terrestrial carbon sink to carbon source, due to temperature-enhanced soil and plant respiration overcoming CO_2-enhanced photosynthesis. This could result in desertification of many world regions as there is predicted to be widespread loss of forests and grasslands, and accelerating warming through a feedback effect. We could also begin to see the collapse of the Amazon rainforest, replacing forest by savannah with enormous consequences for biodiversity and human livelihoods.

(viii)   at around 5°C rise above pre-industrial temperatures (towards the upper end of the estimates in the IPPC's Third Assessment Report) symposium experts predicted a 50% probability of thermo-haline circulation shutdown. Under this scenario the melting of the Greenland ice sheet and the West Antarctic ice sheets may interact with the climate in ways that we have not begun to understand.

On the basis of our best knowledge and understanding today, we can see that the consequences of greenhouse gas emissions rising to beyond 550 ppm (the doubling of concentrations compared with pre-industrial levels) could be very serious indeed.

How climate change will affect our lives

The UK as an example

The UKCIP (UK Climate Impacts Programme) was established in April 1997 to help UK organizations assess and prepare for the impacts of climate change. UKCIP is an independent activity based at the University of Oxford's Environmental Change Institute. It is a highly respected source of impartial and objective analysis and information on climate change as it is likely to impact on the UK. Its report, *Climate Change Scenarios for the United Kingdom,* was published in 2002 (UKCIP, 2002).

More extreme weather events are becoming a feature of the changing climate in the UK. Although no single extreme weather event can be put at the door of climate change per se, events taken together, over time, align well with the kind of trends which the climate change models predict. TV and media coverage of the flooding in Boscastle in 2004, in Carlisle at the beginning of 2005, and of tornadoes in Birmingham in summer 2005 have faded from the media but they are very much a live issue for those people directly affected. UKCIP studies indicate that climate change could have far reaching effects on the UK's environment, economy, and society. Without deep cuts in emissions, average temperatures could rise by about 3°C by 2100 bringing with it more variable and more extreme weather events. It is worth putting this temperature rise in context. By the 2040s or so, on current trends, UKCIP considers that the 'anomolous' summer 2003 temperatures will be the norm; and by the 2080s, the summer 2003 temperatures could be regarded, relatively speaking, as being on the cool side.

Rainfall could increase by as much as 10% over England and Wales and 20% over Scotland by the 2080s. Seasonal changes are expected, with models suggesting that UK winters and autumns will get wetter, and that spring and summer rainfall patterns will change so that the north west of England will be wetter and the south east will be drier. Of course, individual years and groups of years will continue to show considerable variation about this underlying trend. However, the frequency of extreme weather events, such as severe floods, is more likely to increase than decrease. It is less clear at the moment how the frequency of storms and high winds could be affected by climate change.

Some striking evidence of the increase in frequency of climate-change-related events concern the Thames Barrier which was opened in 1983. The last tidal flood of great significance in London occurred in 1928. However, in response to increased awareness of the risk of tidal flooding up the Thames, it was decided to construct the Barrier. At the time of its design in the 1970s, it was expected that it would be used once every few years. In practice, the Thames Barrier (see Fig. 2.12) has been used 6 times a year, on average, over the past 6 years—a clear measure of the increased storm surge levels on the eastern coastline of England.

Developing countries and emerging economies

Developing countries are in the front line so far as the adverse impacts of climate change are concerned. Fig. 2.13 from the IPCC Third Assessment Report shows the results of modelling of global temperature changes over the period 2071–2100 compared with 1961–90.

In his foreword to the report 'Up in Smoke', (IIED, 2004), Dr R. K. Pachauri (Chairman of the IPCC) said:

'The impacts of climate change will fall disproportionately upon developing countries and the poor persons within all countries, thereby exacerbating inequities in health status and access to adequate food, clean water and other resources.'

We have seen examples of extreme weather events and the beginnings of trends—for example, in rainfall—which will be crucial to the future existence of millions of the poorest, most vulnerable people on earth. The report, published in October 2004, describes the plight of poor farmers in the tropical and subtropical areas of the world. They depend on rain-fed agriculture and are barely able to achieve a subsistence level of existence. Variations in precipitation levels, degradation of soil quality, and increased frequency of extreme weather conditions could make the lot of poor peasants far more difficult—even more so than it is today. Some populations will seek to migrate to areas where the environmental conditions are more likely to sustain them. How this migration might impact on neighbouring lands and their peoples is open to question. The significance of climate change on the geo-political stability of many parts of the world is only just beginning to be recognized.

Included in the group of developing countries are some of the fastest growing economies—China, India, Brazil, Mexico, and Russia. They are fuelling their economic development with coal and other fossil fuels. They are emitting carbon dioxide at a rate which is exacerbating the upward trend in atmospheric carbon

(a)

(b)

Annual Closure of Barrier

Number of Thames Barrier closures against tidal surgess, 1983–2002

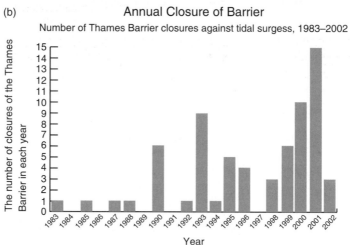

Figure 2.12   (a) The Thames Barrier, downstream of London, which is closed during tidal surges. (b) Graph showing number of Thames Barrier closures over the period 1983–2002. Source: the Environment Agency.

dioxide concentrations. On current trends, China, for example, is set to overtake the US as the largest global carbon dioxide emitter within the next 10 years. By 2030, it is estimated that coal plants in developing countries could produce more carbon dioxide emissions than the entire power sector of OECD countries does now.

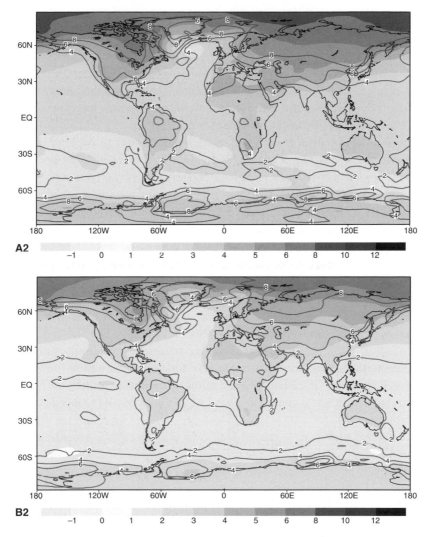

Figure 2.13 Modelling the annual mean change of the temperature in °C (shading) and its range (contours of constant temperature) (Unit: °C) for the period 2071 to 2100 relative to the period 1961 to 1990. Source: IPPC, 2001.

Taking action

It is clear that the climate is changing—for all living things on earth. We are moving into uncharted territory so far as the impacts on flora and fauna, people and businesses, and countries and global regions are concerned. We can, and should, take action now to move to a low carbon global economy. If we start now, we stand a chance of achieving the huge cuts in greenhouse gases required over the decades ahead and, hopefully, of avoiding the worst impacts of climate change in the centuries ahead. If we take insufficient action, future generations will find it much more difficult to halt, and subsequently reduce, atmospheric greenhouse gas levels and there will be significant changes to their environments, to food growing areas of the world, and to patterns of human migration.

We can all take action to cut out energy waste, buy energy efficient equipment and products, and reduce the amount of high-carbon energy we buy. However, in order to make a material impact, individual actions need to be encouraged and consolidated into national, and global, action. Creating effective policies and measures to mediate the transition to a low-carbon economy is the role and responsibility of Governments—all Governments. In developed countries it means governments working with business, the public sector, and individuals, to set up programmes that encourage and support action and investment to improve energy efficiency and to reduce the carbon intensity of economic activity.

The scale of global activity required to halt and reduce greenhouse gas emissions to around 60% by the middle of this century is huge. A sense of the scale involved can be seen from the work of Robert Socolow and his colleagues at Princeton University, USA (Socolow, 2004). They have developed the concept of *stabilization wedges*. A wedge represents an activity that reduces emissions to the atmosphere, starting at zero today and increasing linearly until it delivers around 1 Gt carbon/year of reduced carbon emissions in 50 years' time. It thus represents a cumulative total of 25 Gt carbon of reduced emissions over 50 years.

A number of nations have started their own climate change programmes. The UK Government's Climate Change programme, launched in 2000, is an example of early action at the national level designed to show leadership, achieve Kyoto obligations, and make progress towards a domestic goal of a 20% reduction in carbon dioxide emissions by 2010 based on 1990 levels (DETR, 2000). The UK Government's Energy White Paper of 2003 confirmed that climate change is central to its environment and energy policies (DTI, 2003). It accepted the Royal Commission on Environmental Pollution's recommendation, in its 2000 report on the changing climate, that the UK should put itself on a path towards a reduction in carbon dioxide emissions of some 60% from current levels by about 2050 (Royal Commission, 2000). For business, the public sector, and individual consumers it means accepting and being part of, the transition to a low-carbon economy: cutting out energy waste; investing in and preferentially purchasing lower carbon intensity heat and power; and making choices which reduce the carbon footprint of economic, social, and leisure activities. Much can be achieved with today's technology and know-how. Therefore, a considerable part of the UK's Climate Change programme has been focused on creating incentives for investment in energy efficiency measures and the deployment of commercial sources of low-carbon electricity.

Raising awareness about climate change in the respective consuming sectors of the economy is an essential pre-requisite to taking action. Individual consumers have a mixed perception of climate change and their own roles and responsibilities to reduce carbon dioxide emissions. Some, not many, are aware and have adjusted their purchasing decisions accordingly. The majority, however, have demonstrated by their purchasing behaviour that they: (a) do not understand their impact on carbon dioxide emissions; (b) or do not understand how to reduce their carbon dioxide emissions; (c) or do not care. There is a huge challenge to raise consumer awareness, change behaviours, and stimulate action. The same, generally speaking, applies to the business community, though like domestic consumers there are leaders and laggards. Corporate business, thought of by some as being in the 'problem' rather than the 'solution' camp, is showing that not only is it aware of climate change and the impacts it is beginning to have, but is also keen to be pro-active. Thirteen major UK and international companies offered to work in partnership with the Government towards strengthening domestic and international progress on reducing greenhouse gas emissions. A key quote from their letter to Prime Minister Tony Blair in May 2005 (Corporate Leaders Group on Climate Change, 2005) is: 'Enabling a low-carbon future should be a strategic business objective for our companies and UK plc as a whole'. Canadian corporate leaders wrote a similar letter to Canadian Prime Minister Paul Martin in November 2005 (Canadian Executive Forum on Climate Change, 2005). Their opening paragraph said:

'As corporate leaders representing a broad cross-section of the Canadian economy, we believe that all governments, corporations, consumers and citizens have responsibilities under the Kyoto Protocol and that the world must act urgently to stabilize the accumulation of greenhouse gases in the atmosphere and minimize the global impacts of climate change.'

Making the transition to a low-carbon future will require not only widespread systematic deployment of the best of today's technology, but also a concerted effort to innovate, develop, and commercialize new low carbon and energy efficient technologies. The huge reductions in carbon dioxide emissions that are required will not be achieved without developing and deploying energy efficiency and low carbon technologies at scale in existing and emerging economies, and as part of a global endeavour. What is required is no less than a global approach to developing and commercializing a new generation of low carbon technologies and products. The power and innovation of multi-national energy companies, technology companies, and the investor communities need to be harnessed and encouraged to achieve the low-carbon economy goal. This will not happen unless the right business environment can be created. Helping the market to deliver huge reductions in carbon dioxide emissions at scale and on the desired timescales, whilst maintaining and improving economic and social wellbeing, requires intervention by and cooperation between governments, and between governments and business. Making the transition from fossil fuel, high-carbon investment to low-carbon, sustainable (environmentally, socially, and commercially) investment, requires a common sense of purpose and recognition that it makes business sense to do so. Accelerating the commercialization of new and emerging low carbon technologies is crucial to achieving a timely transition to a low carbon economy.

Technological and policy innovation

Over the last 100 years or so, there have been many examples of technological innovation: 'cats eyes' on roads and motorways worldwide; television; mobile phones; etc. These are very different technologies but, like many global scale innovations, both have changed the everyday lives of millions of people. Energy supplying and consuming technologies pervade the lives of billions of people in the developed, and increasingly in the developing, world. Making sure there are efficient, affordable and reliable low-carbon solutions to our energy needs is part and parcel of the transition to a low-carbon economy. However, that transition will not happen overnight. It took 100 years from the earliest heavier-than-air–machine, skimming a mile or so over that windswept North American landscape in 1903, to the modern aeroplane fleets and, importantly, the international airline infrastructure, to mature to a commercial, affordable, and (in the main) efficient service. Furthermore, the pace of development was accelerated by two world wars which gave a huge impetus to aircraft and engine design innovation.

The energy sector is one where price, demand, and the strength—or otherwise—of government intervention impact significantly on the pace of innovation and commercialization. 2005 was the year in which the $60/barrel of oil came into existence. Whether this was simply a short term reaction to current political and economic uncertainties, or a trend-setting market reaction to future supply availability in a world with rapidly growing demands for energy, nobody really knows for sure. Short term price fluctuations have been a characteristic of world energy markets for decades. However, when price fluctuations become an upward trend, it drives the search for new energy sources and new energy technologies—for example, the exploration and production of oil from reserves which are increasingly more costly to exploit but which are, nevertheless, profitable investments.

Another driver for new and emerging energy technologies and low-carbon technology innovation is the finite nature of our oil and gas reserves. Some independent oil consultants suggest that at present rates of consumption and discovery, world oil production will peak between 2015 and 2020. Predictions of production peaks and their timings are, however, fraught with dispute and controversy. What is not in dispute is that demand for energy is rising—driven by the emerging economies. These factors are strong economic drivers to explore for more reserves and to seek alternatives to oil for energy supplies.

Many countries are looking at which new and emerging low carbon technologies make climate change and business sense to them. In 2002, as part of a UK government-commissioned study to examine the long-term challenges for energy policy (Performance and Innovation Unit, 2002), the government's Chief Scientific Adviser, Sir David King, led a review of energy research in the UK. The conclusion (Energy Research Review Group, 2002) was that there were six areas of research in particular where there was significant headroom between where the technology is today and where it could be if more research, development and demonstration (RD&D) were supported. These included: carbon capture and storage, which might enable us to continue to burn fossil fuels by collecting the carbon dioxide and sequestering it safely in suitable geological formations for the long term; energy efficiency gains across the energy consuming sectors; hydrogen production, distribution, use, and storage; nuclear fission and radioactive waste management; materials for fusion reactors; and renewable energy technologies including, in particular, solar photovoltaics, wave and tidal. To coordinate energy R&D in the UK, the government set up the UK Energy Research Centre (UKERC) in 2005, and in his Budget 2006, the Chancellor announced the formation of what is now called the Energy Technologies Institute– a public-private partnership to support RD&D into energy and low-carbon technologies.

Energy markets, and their shaping by policies, priorities, and events, are also crucial factors. In the UK, for example, the energy generators, network operators and suppliers are part of a privatized, regulated industry in what is one of the most liberalized energy markets in the world. The degree of regulation and liberalization, coupled with government policies of the day and public opinion, are the main determinants of the climate for energy technology innovation. Thus, for example, increasing the percentage of low-carbon electricity from the UK supply system and decentralized sources, depends not only on government policies on clean coal, carbon capture and storage, and nuclear power, but also, in part, on the extent to which there are incentives for grid owners to invest in the necessary new transmission and distribution lines to bring low-carbon power to consumers.

At the heart of the issue are questions of the type raised by the UK Energy regulator (Ofgem), government, and informed commentators:

(i) whether, and to what extent, a liberalized market can provide sufficient stimulus for low-carbon technology (including the infrastructure) innovation and hence deliver a low-carbon economy; and

(ii) if government intervention is needed, what form should it take, who needs to be encouraged, and for how long should the intervention measures operate.

Multinational energy technology companies will respond to the different market opportunities according to their overall attractiveness and strategic value. They have the capacity and the market position to capitalize on innovation— either 'home-grown' or acquired from bought-out, smaller companies. They will dominate the development of standardized energy technologies. The race to develop and deliver standardized, communicating products to global scale markets (sometimes regulated, sometimes less so) will be a driving strategy. Countries like the UK, representing only a few percent of global demand, will not, individually, present large enough markets to influence the nature and direction of innovation for the bulk energy supply technologies of the future. Conversely, making one-off 'specials' to fit a particular national standard or specification will be supplied but at premium rates.

Much has been written about technological innovation and the roles of the market and the State. Academic papers, theories, and models abound. One is the work carried out by the Environmental Policy and Management Group (EPMG) at Imperial College, London. Their research focuses on how to develop better policy processes to promote sustainable innovation for achieving social, environmental, and economic goals.

In the following chapters, there are descriptions of a range of low carbon technologies and concepts—from nuclear fusion through carbon capture and long term storage, to novel photovoltaic devices. Energy efficiency technologies, designs, and products are also covered. So often, reducing demand and cutting out energy waste is either forgotten or ignored. And yet it is an important step in the process of decarbonizing economies, improving resource productivity, and saving money. The energy supply side is regarded by many as where the exciting action is—for example, smart science and engineering, market-driven technological innovation, large construction projects, etc. Energy efficiency is regarded as boring by some, ineffective by others. In fact, the technological, attitudinal, and behavioural challenges on the demand side are just as challenging as they are on the supply side. Energy efficiency savings are a reality not a myth. Through procurement decisions based on whole life costs and a carbon life cycle analysis, backed up by a higher standard of energy management, it is possible to make significant energy savings using today's know-how. Savings in demand of the order of 20–30% are not considered to be unreasonable (Performance and Innovation Unit, 2002). There are opportunities everywhere in industry, business, transport, and the public and domestic sectors, with building and transport providing particularly important opportunities.

Summary and conclusions

In this chapter, the greenhouse effect has been described, and the growing body of scientific evidence supporting the existence of anthropogenically forced climate change has been summarized. The link between global temperature rises and greenhouse gas emissions (particularly carbon dioxide from the combustion of fossil fuels) has, for most informed people, been demonstrated beyond reasonable doubt. Examples from the growing body of weather phenomena have been quoted as being indicative, not predictive, of the kind of weather events which we might experience more of in the decades to come. The pace of anthropogenically forced climate change is faster now than we thought just a few years ago. What we are seeing may be 'benign' compared with the frequency and intensity of weather events to come. There are still uncertainties to be explored. However, as computing power develops and more data is gathered it will be possible to develop and test the ever-increasingly complex scientific models required to improve our understanding of climate change and its impacts.

The general response from many developed countries to anthropogenically forced climate change has been to ratify the Kyoto Protocol—a first global step to reducing carbon dioxide emissions and tackling climate change. Other countries, including the US, China, and Australia, are exploring other ways to collaborate on cleaner and low-carbon technologies. In the UK, the government has a goal to reduce carbon dioxide emissions by 20% by 2010 on 1990 levels. It also has a longer term aspiration to reduce them by some 60% by around 2050 based on 1997 levels. The scale of reduction is regarded as not just a UK goal, but a global one. They are not targets or end games in their own right. They are markers on the road to a low-carbon economy. What concentration of greenhouse gases in the global atmosphere is 'safe' is, to be honest, unknown.

Developing countries are at a stage when energy consumption, economic growth (from a relatively low base), and standards of living are inextricably linked. Using the cheapest source of energy is the only option available to them, and this often means using indigenous fossil fuels, thereby forcing carbon dioxide emissions upwards. Paradoxically, they are also at the 'sharp end' of extreme weather events associated with anthropogenically forced climate change. Making the global transition to a low-carbon economy is therefore particularly important for developing countries.

Developed countries are beginning to create energy choices that include low carbon technologies. However, the nature, direction, and pace of innovation of technologies for historically low-risk, relatively low-return utilities is heavily dependent on the degree of intervention from governments and the regulatory authorities (where energy markets have been liberalized).

No single technology or concept will achieve the global goal of tackling anthropogenic climate change. No single policy approach will yield success. The solutions for developed countries seeking to decarbonize their established economies from an established fossil fuel base, will be different from those of developing countries seeking to raise standards of living, gain a foothold in the global economy, and take action to avoid the worst impacts of the extreme weather events and climate trends to which they are particularly vulnerable. Both mitigation and adaptation strategies will be required. On mitigation, we know that today's technologies will help us make a start towards decarbonizing our economies, but they will not be sufficient. Innovation to develop and deploy a new generation of low-carbon supply and demand side technologies will be essential. Innovation—and commercialization—will not happen at the pace and on the scale required, unless governments and markets work to create a framework which provides incentives to stimulate the necessary investment in, and to expand the necessary customer base for, low carbon technologies and products.

Postscript

Since this chapter was written in the first half of 2006, the UK Government's Energy Review and the Stern Review on the Economics of Climate Change have been published. The Energy Review, published in July 2006, put forward proposals for tackling the huge energy and climate change challenges we face, at home and abroad (Energy Review, 2006). The Stern Review, the first major review of the economic implications of climate change, was published in October 2006. This review painted an economist's picture of the costs of taking action to tackle climate change and the far bigger costs if we fail, globally and collectively, to take action in the next decade. In summary, the Stern Review concluded that:

'The costs of stabilising the climate are significant but manageable; delay would be dangerous and much more costly. Action on climate change is required across all countries, and it need not cap the aspirations for growth of rich or poor countries.'

In these few words, the value proposition for humanity is clearly expressed. Start now to invest in the transition to a low-carbon economy and we stand a better chance of a viable future for tomorrow's generations. Prevaricate and we face an increasingly higher probability that climate change will become irreversible and change the course of our world forever. The choice, surely, is unequivocal.

Resources and further information

Canadian Executive Forum on Climate Change, 2005. Open letter to the Canadian Prime Minister from the Executive Forum on Climate Change Call to Action on Climate Protection, Energy and Sustainable Development, dated November 2005.

Corporate Leaders' Group on Climate Change, 2005. Open letter to the Prime Minister from HRH The Prince of Wales's Business & the Environment Programme Corporate Leaders' Group on Climate Change, dated 27 May 2005.

Department of the Environment, Transport and the Regions, (DETR) 2002. *Climate Change – The UK Programme*, Department of the Environment, Transport and the Regions, London.

DTI, 2003. *The Government's Energy White Paper of 2003*: Our energy future creating a low-carbon economy. Available from www.dti.gov.uk/publications

Energy Review Report, 2006, *The Energy Challenge*, Department of Trade and Industry, published July 2006.

Environmental Policy and Management Group, Imperial College. Transforming policy processes to support sustainable innovation: some guiding principles: www.imperial.ac.uk/environmentalscience/research/epmg/EPMGFrontpage.html www.iccept.ic.ac.uk; and on the Economic and Social Research Council's sustainable technologies programme at: www.sustainabletechnologies.ac.uk.

Foxon et al., 2005. Transforming policy processes to support sustainable innovation: some guiding principles. Environmental Policy and Management Group, Imperial College, www.imperial.ac.uk/environmentalscience/research/epmg/EPMGFrontpage.html www.iccept.ic.ac.uk; and on the Economic and Social Research Council's sustainable technologies programme at: www.sustainabletechnologies.ac.uk.

International Institute for Environment and Development (IIED) and the New Economics Foundation, 2004. *Up in Smoke: threats from and response to the impacts of global warming on human development*. www.neweconomics.org.

The Intergovernmental Panel on Climate Change (IPCC) 2001. *Third Assessment Report*. Available from the IPCC website www.ipcc.ch

McManus, 2004, Glacial cycles of the past 800,000 years. *Nature, 429*, 623–28.

Royal Commission on Environmental Pollution's 22nd report, 2000. *The Changing Climate* www.rcep.org.uk/newenergy.html

Schaer et al. 2004. Localised temperature anomaly–summer 2003. *Nature, 427*, 332–36.

Schellnuber, J. (ed) 2005. Avoiding Dangerous climate change–A Scientific Symposium on Stabilisation of Greenhouse Gases. 1–3 February, 2005 at the Hadley Centre for Climate Prediction and Research, the Met Office, Exeter, United Kingdom.

Socolow, J. 2004. Stabilization wedges: solving the climate problem for the next 50 years with current technologies. *Science, 305*. www.sciencemag.org

The Performance and Innovation Unit's Energy Review– a report to Government published in February 2002.

The Chief Scientific Adviser's Energy Research Review Group review of energy R&D– 2002 available from www.ost.gov.uk

The Stern Review, 2006, *The Economics of Climate Change.*

UKCIP, 2002. The UK Climate Impacts Programme Climate Change Scenarios for the United Kingdom, 2002. *The UKCIP02 Scientific Report April* 2002. Available from: www.ukcips.org.uk

Some further UK focused reading and sources

The Climate Change Challenge 1: Scientific evidence and implications–a Carbon Trust publication available from www.thecarbontrust.co.uk

Climate Change and the Greenhouse Effect–a briefing from the Hadley Centre, Met. Office: www.metoffice.gov.uk/research/hadleycentre/. Hadley Centre for Climate Prediction and Research–www.hadleycentre.com. The Hadley Centre is the UK Government's centre for research into the science of climate change.

Tyndall Centre for Climate Change Research-www.tyndall.ac.uk. The Tyndall Centre, funded by three UK research councils, is the national centre for trans-disciplinary research on climate change.

The UK Energy Research Centre–www.ukerc.ac.uk

The author

Dr David Vincent is Technology Director of the Carbon Trust, a private company set up by the UK Government in response to the threat of climate change to accelerate the transition to a low-carbon economy. He trained as a chemical physicist and, motivated by the energy crises of the early 1970s, decided to pursue a career in the energy field. He has held a variety of posts in UK Government Departments focusing on energy efficiency and low-carbon technology RD&D, and energy/climate change policy development. His interests include low-carbon technology innovation and commercialization and the interaction between Governments and markets to drive low-carbon choice and investment.

3. *Geothermal energy*

Tony Batchelor and Robin Curtis

Introduction

The term 'geothermal energy' describes all forms of heat stored within the Earth. The energy is emitted from the core, mantle, and crust, with a large proportion coming from nuclear reactions in the mantle and crust. It is estimated that the total heat content of the Earth, above an assumed average surface temperature of 15°C, is of the order of 12.6×10^{24} MJ, with the crust storing 5.4×10^{21} MJ (Armstead, 1983). Based on the simple principle that the 'deeper you go the hotter it gets', geothermal energy is continuously available anywhere on the planet. The average geothermal gradient is about 2.5–3°C per 100 metres but this figure varies considerably; it is greatest at the edges of the tectonic plates and over hot spots–where much higher temperature gradients are present and where electricity generation from geothermal energy has been developed since 1904.

Geothermal energy is traditionally divided into high, medium, and low temperature resources. Typically, temperatures in excess of 150°C can be used for electricity generation and process applications. Medium temperature resources in the range 40°C to 150°C form the basis for 'direct use' i.e. heating only, applications such as space heating, absorption cooling, bathing (balneology),

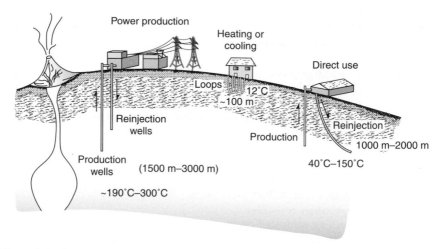

Figure 3.1   Cartoon showing the basic principles of extracting geothermal energy.

process industry, horticulture, and aquaculture. The low-temperature resources obtainable at shallow depth, up to 100–300 metres below ground surface, are tapped with heat pumps to deliver heating, cooling, and hot water to buildings.

The principles of extracting geothermal energy, in applications ranging from large scale electrical power plants to smallscale domestic heating, are illustrated in Fig. 3.1.

Geothermal energy can be utilized over a temperature range from a few degrees to several hundred degrees, even at super critical temperatures. The high temperature resources, at depth, are typically 'mined' and are depleted over a localized area by extracting the *in situ* groundwaters and, possibly, re-injecting more water to replenish the fluids and extract more heat. Although natural thermal recovery occurs, this does not happen on an economically useful timescale. On the other hand, the low temperature resources can be designed to be truly renewable, in the sense that the annual rate of extraction can be designed to be matched by the rate of recovery.

Where warm water emerges naturally at the Earth's surface, man has probably used geothermal energy since prehistoric times. On a commercial scale, electricity generation using geothermal energy is now over 100 years old, with 24 countries having plants on line. Direct use and heat pump applications are recorded for over 70 countries.

It is not generally appreciated that geothermal energy currently ranks fourth in the league table of 'alternative' energy sources in terms of energy delivered—after biomass, hydropower and, very recently, wind (REN21, 2005).

Geothermal energy has five key characteristics that can deliver important benefits as an energy source supplying heat:

- It provides a very large resource base, readily available in one form or another in all areas of the world.

- It is a reliable and continuous source of energy and can provide base load electricity, heating, cooling, and hot water in the right circumstances. There is no intermittent nature to the resource.

- The technology is maturing and geothermal energy can be economically competitive as long as applications are designed correctly and are matched to geological conditions.

- It can leverage the role of other forms of renewable or carbon-free electricity by factors of three to four when used with heat pumps in heating and air conditioning applications; i.e. one unit of electrical power can deliver four units of carbon free heat.

- It is already accepted in various sectors of the market place both for investment and operations, although the technologies are not yet understood by a wide audience.

In the context of this book, two additional features of geothermal energy should be highlighted.

- If the world moves towards a hydrogen economy, there will be a need for non-fossil fuel sources to provide the energy for hydrogen production from water. Given that geothermal energy is ideal for operation at a steady base load, but cannot in itself be transported long distances, it can form an excellent basis for hydrogen production—with the hydrogen itself then being transported to point of use. This already forms part of the Icelandic proposals to become the first hydrogen based economy in the world (Sigfusson, 2003).

- At the low-temperature end of the spectrum, the fortuitous role of geothermal direct use or geothermal heat pumps in matching the temperature requirements for the heating and cooling of buildings, presents one of the few currently available options for eliminating the use of fossil fuels (and their resulting carbon emissions) as the dominant energy source for providing thermal comfort in buildings.

The reader who wishes to go beyond the discussion presented here is referred to the online article by Dickson and Fanelli (see web resources below) which is drawn from their UNESCO publication (Dickson and Fanelli, 2003), or the excellent review paper in *Renewable and Sustainable Energy Reviews* (Barbier, 1997).

High-temperature resources

Fig. 3.2 shows the 24 countries that had established installations for geothermal electrical power generation in 2005. Typically, the high-temperature geothermal industry uses geothermal fluids from the ground at 200°C to 280°C from wells 1,500 to 2,500 metres deep. This type of resource is only found in regions with active volcanism and tectonic events on major plate and fault boundaries. The highest grade of geothermal energy is dry steam; it only occurs in rare and geographically limited locations where the *in situ* fluids can exist as steam (for example, the Geysers, California and Larderello, Italy) and can be fed directly to turbines. More commonly, however, the fluids are held in the rock as hot, pressurized liquids that can convert spontaneously to steam at the surface to drive turbines. In lower temperature systems and systems with difficult chemistry conditions, the geothermal fluids are fed to heat exchangers where a secondary circuit heats a closed, clean fluid that is used to power a turbine or rotary expander. These latter systems will become more important as the use of lower temperature resources is increased (DiPippo, 2005).

Location of Geothermal Power Sites, 2005

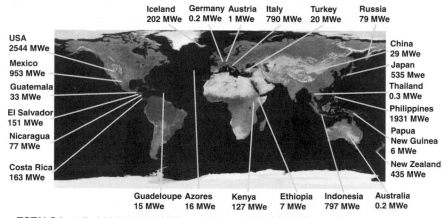

Iceland	Germany	Austria	Italy	Turkey	Russia
202 MWe	0.2 MWe	1 MWe	790 MWe	20 MWe	79 MWe

USA
2544 MWe

Mexico
953 MWe

Guatemala
33 MWe

El Salvador
151 MWe

Nicaragua
77 MWe

Costa Rica
163 MWe

China
29 MWe

Japan
535 Mwe

Thailand
0.3 MWe

Philippines
1931 MWe

Papua
New Guinea
6 MWe

New Zealand
435 MWe

Guadeloupe	Azores	Kenya	Ethiopia	Indonesia	Australia
15 MWe	16 MWe	127 MWe	7 MWe	797 MWe	0.2 MWe

TOTALS Installed 2000: 7,974 MWe, and 2004: 8,912 MWe (Generated 56,798 GWh/y)

© Tony Bachelor, 2005

Figure 3.2 Map showing location of the principal regions of high-temperature geothermal power production. Note that MWe denotes Mega watts of electricity as distinct from MW$_t$ (Mega watts of heat energy).

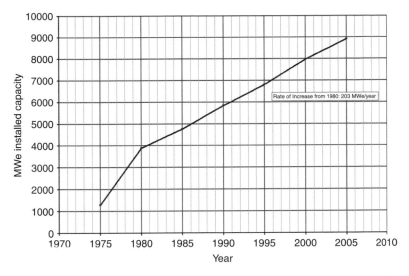

Figure 3.3 Growth in installed capacity of electricity generation from geothermal energy.

In 2004, the reported installed generation capacity was 8,912 MW$_e$, generating 56,798 GWh of electricity per annum—an average availability of 72%. Fig. 3.3 shows the sustained growth rate in installed capacity of 203 MW$_e$ year, a steady capital expenditure of close to $1 billion per annum.

The overall fraction of geothermal power generation compared to the world total power generation is currently 0.4% and the goal of the geothermal community is to raise that fraction to 1% by 2010. However, the apparently small global fraction masks the local importance of geothermal power production. Fig. 3.4 shows the importance of indigenous geothermal power to certain developing countries; it is worth noting that geothermal power is economically competitive with hydropower (for example, Central America and New Zealand) under the right circumstances.

A detailed worldwide assessment of the state of geothermal electricity production, reviewed following the 2005 World Geothermal Congress is available (Bertani, 2005).

Hot dry rock or enhanced geothermal systems

Given the concept of 'the deeper you go the hotter it gets', the 'blue skies' aspiration of the geothermal industry has been the Hot Dry Rock (HDR) concept— now also referred to as Enhanced Geothermal Systems. With the resource available everywhere, the goal is to drill to the required depth (temperature)

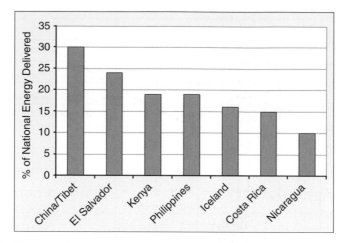

Figure 3.4   Utilization of geothermal energy in Iceland and some developing countries.

to engineer a large permeable heat transfer surface between one or more wells, to inject cold water at depth and to recover it at temperatures suitable for the production of electricity. Three major research projects have been undertaken in this technology, at Los Alamos, USA in the 1970s and 1980s, in Cornwall in the UK in the 1980s and 1990s, and currently at Soultz in France. Other research and development work has also been undertaken in France, Sweden, Germany, and Japan. Pro-commercial projects are now underway in Switzerland and Australia.

The HDR concept (see Fig. 3.5) is not new and was first described by Parsons in 1904, advancing suggestions from a Royal Society committee. The great attraction is that all the energy requirements of most communities can be met by drilling to 5,000–7,000 metres deep; the problem is creating the interlinking system between wells with sufficient heat transfer area at these great depths. Since 1970, there has been approximately 35 years of work (representing several thousand man years of effort) and about $500 million spent worldwide in advancing this technology.

An internal report from the European Commission has suggested that 15% of Europe's power could come from this deep technology. Cornwall in the UK is one of the best potential locations in Northern Europe outside the volcanic regions. Parsons, when he described the basic process, said it would take 85 years before anybody took his ideas seriously—perhaps that time is arriving. There is also interest in using the HDR concept at shallower depths, purely for the production of heat, and several European countries have projects underway or actually in operation. An in depth review of this technology was recently presented at The Royal Society, (Tester *et al.*, 2006)

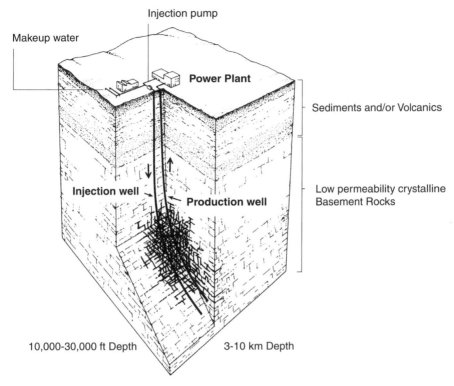

Injection pump

Makeup water

Power Plant

Sediments and/or Volcanics

Injection well

Production well

Low permeability crystalline
Basement Rocks

10,000-30,000 ft Depth

3-10 km Depth

Figure 3.5 A hot dry rock (HDR) geothermal installation.

Another area of advanced work for the geothermal industry is the exploitation of supercritical resources. At some locations it is possible to access fluids at temperatures and pressures in the super critical regime (for water). These conditions have been identified in Japan and Italy and are now being exploited in Iceland (Fridleifsson and Elders, 2005). Provided that the borehole pressures and temperatures can be handled safely, the specific power output of this type of resource is considerably higher than conventional power production wells.

Medium-temperature resources

More than 70 countries have installations taking hot water directly from the ground at 40°C to 150°C, over three times the number with high-temperature applications. The top four countries in terms of direct use, China, United States, Iceland, and Turkey, account for 68% of the geothermal energy used directly as heat. As the required source temperature diminishes, so the geographic constraints relax. For example, even in the UK (BGS, 1986) there is one famous geothermal spa complex at Bath, and one geothermally fed district heating system at Southampton. There is potential for a few more locations; but the economics are generally unattractive at current gas prices (Batchelor *et al.*, 2005). Fig. 3.6 shows the distribution of types of direct uses without heat pumps; worldwide,

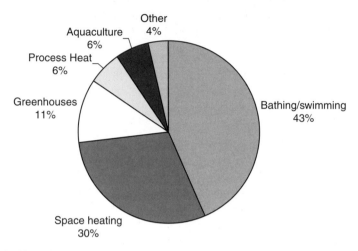

Figure 3.6 Natural uses of geothermal energy worldwide (without heat pumps).

Figure 3.7 Two wells in Hungary supplying heating water to a housing complex.

these applications utilize approximately 48,700 GWh (thermal), saving at least 10 million tonnes of carbon dioxide per year compared to burning gas.

Fig. 3.7 shows an example of a direct use application in Hungary—in this case two wells supply medium temperature geothermal water to a housing complex.

For a detailed assessment of direct use utilization the reader is referred to the recently published worldwide review undertaken following the 2005 World Geothermal Congress (Lund, Freeston, and Boyd, 2005).

Geothermal energy as a sideline of oil and gas industries

Whereas deliberate efforts to extract geothermal energy often present challenges for commercial development, the oil and gas industries achieve this as a sideline. One major oil company has recently been involved in evaluating the feasibility of using the associated production of water, at moderate temperature, to offset power requirements for processing. Technically it appears feasible; a US university study has also looked at the utilization of produced water (Mckenna and Blackwell, 2005). Currently, the key contribution that the oil and gas business makes to geothermal energy is exploration. The oil companies, of course, are less than delighted when they encounter large quantities of hot water! The HDR site at Soultz in France was discovered by drilling beneath an oil field (Pechelbron) into a large hot water resource.

Work is underway to investigate the use of the higher temperature produced fluids from oil and gas operations, with a UK designed, high-efficiency binary expander (Stosic *et al.*, 2003) to generate electricity locally for certain oil and gas production facilities. This same expander appears to offer the opportunity for significant growth in power generation from medium temperature resources.

Low-temperature systems with heat pumps

As the requirements for specific ground temperatures for particular applications are relaxed, the areas suitable for viable geothermal utilization expand rapidly. The two previous categories of resources are generally associated with large, semi-centralized systems because of the need to use deep, capital-intensive boreholes in order to reach high enough temperatures. By contrast, low ground-temperature systems used with heat pumps are suitable for small scale, even domestic, systems and can be used anywhere to heat, to cool, and to provide domestic hot water. The technology is not new; it has roots back to William Thomson, (Lord Kelvin) in 1852 and work by a Royal Society committee (Royal Society, 1869).

Worldwide, the installed base of geothermal heat pumps now exceeds 15,750 MW$_t$, a sevenfold increase of capacity in 10 years; almost all these installations are in North America and Europe, estimated as 1.3 million $12\,kW_t$ equivalent units (thought to be about two million actual installations, Systemes Solaires, 2005). Geothermal heat pumps are saving over four million tons of carbon dioxide a year (provided they displace gas as the heating fuel).

The geothermal sources are either open-loop systems using the ground water directly through an evaporator heat exchanger, or closed-loop systems using a water based antifreeze mixture circulating through sealed pipework in boreholes or trenches. One variant on open-loop systems that is becoming attractive is the use of abandoned mines to form the subsurface 'heat exchanger'. An entire conference has recently been devoted to this potential energy resource (see Malolepszy, 2001). For example, in the UK, large projects are being considered for the former Monktonhall Colliery in Scotland and in the Camborne and Pool Urban Regeneration area of Cornwall using abandoned tin and copper mines clustered around the old Dolcoath Mine, once the deepest and hottest mine in the world.

Closed-loop installations allow the technology to be used anywhere; for example, in Europe they range from the smallest housing unit using a $4\,kW_t$ heat pump to a $9,000\,kW_t$ heating system in Norway. Switzerland has the highest density of units at over 1 system per $2\,km^2$ and more than 30,000 operating systems, with 3,000 units per year being installed. In contrast, the UK has about 550 systems (early 2005) with some 100 new systems under construction in early 2005.

Development of reliable heat pumps and the availability of high density polyethylene pipework have been key factors in making the technology applicable in a wide variety of situations. A general lack of understanding of the effective use of the systems by architects, building services engineers, and main contractors is perceived as a significant impediment to current growth. The reader is referred to the chapter by Dell and Egger (Chapter 12) for an analysis of the possibilities which have been implemented in Austria.

A typical European $6\,kW_t$ system on a newly built, $100\,m^2$ house will use less than $3,500\,kWh$ of electricity for the heat pump and circulation system to provide about $13,000\,kW_th$ to heat the house and domestic hot water. For one house, this will save 1.5 tonnes of carbon dioxide compared to gas and 3.6 tonnes compared to oil each year (based on UK figures). An added advantage of these systems is that they can incorporate reversible heat pumps to provide cooling at very little additional cost.

An up to date worldwide assessment of geothermal heat pumps is provided by Lund, Freeston and Boyd (2005), with a discussion of worldwide application of the technology summarized in a paper for the 2005 World Geothermal Congress (Curtis *et al.*, 2005).

Potential for future growth

Low-temperature systems with heat pumps will have the biggest impact in the near future on the growth of utilization of geothermal energy. The growth is unlimited: there is no reason why every new building with a normal load profile would not benefit from such systems. Some new buildings are not suitable for geothermal energy, for example those with loads that peak over short intervals, but have otherwise low occupancy, for example religious meeting facilities.

A limitation on the rate of growth for all other forms of geothermal system is the requirement for pre-feasibility studies to determine 'is it hot enough?', 'can it flow at a high enough rate?' and 'is there enough of it?'. Even ground-source heat-pump systems need to be capable of being dug and drilled without undue cost; fortunately only very few sites present insuperable problems. This exploration phase or 'geological risk' assessment prior to establishing a major project is usually funded by national or supra-national organizations. There is a steady flow of long-term power generation projects developing in East Africa, South East Asia and Central/South America that will eventually support over 29,000 MW$_e$ of future systems in these areas (Stefansson, 1998).

Two main studies have examined future growth prospects. The first study which looked at 'proven resources' and existing markets (Reed *et al.*, 1999) concluded that a ten-fold expansion, i.e. a 100 GW$_e$ increase in power production, was feasible. The second report, by Stefansson, has looked further at undiscovered resources and the ultimate size of high-temperature systems (Stefansson, 2000). The analyses show that worldwide geothermal use could increase by a factor of more than 100, limited only by the impact of energy prices and environmental considerations. When the original paper was written, oil was trading at \$10–20 per barrel and the USA was rejecting the concept of significant carbon emission reduction; only 0.2% of Stefansson's total resource is in current use for power generation.

For direct use Stefansson has estimated that the potential useable resources could provide 392,000 TWh/year, or more than three times the annual energy consumption today. The essential point is that the availability of the resource is not the limiting constraint.

There seems to be no limit on the application of heat pumps other than the capability to provide awareness, installation, support, training, and design standards at the rate required for significant growth. Consider a situation with 50 MW$_e$ of firm renewable power from, say, hydropower or geothermal sources driving modern heat pump systems. Such a system would deliver up to 200 MW$_t$ of renewable energy at the points of use with much higher savings of emissions compared to those from the electricity component alone.

Conclusions

Geothermal energy is one potential source of future energy supply that is already established and provides a significant fraction of the renewable energy worldwide. It has the potential for further substantial growth, making a large contribution to the reduction of emissions, particularly in developing countries.

The resource is available in one form or another in every country and can provide base load electricity and heat on a continuous and sustained basis. Universities are already delivering specialist training courses under a UN programme in Iceland, Italy, and New Zealand. Annual training and awareness programmes in direct use are provided through the very successful 'Geothermal Days' programme, supported by the International Geothermal Association. The geothermal community is doing its utmost to raise the profile and awareness of the technologies and to promote the consideration of geothermal energy at the earliest phases of a project.

The future is very bright and secure for geothermal energy; the heat from the Earth will have a major role to play in the era 'beyond oil'.

Acknowledgements

Two main sources of data were used for the preparation of this chapter. They were the principal summaries for the International Geothermal Association, World Geothermal Congress 2005 held in Turkey; one paper covers Power Generation (Bertani WGC, 2005) and the other Direct Uses (Lund *et al.*, WGC 2005). The geothermal community is quite small and many colleagues in GeoScience/EarthEnergy Ltd and in the International Geothermal Association have provided input. Particular assistance was provided by Peter Ledingham and John Garnish. The views expressed in this paper are entirely our own and are not necessarily endorsed by any organization.

Resources and further information

Armstead, H. C. H. 1983. *Geothermal Energy*, E. & F. N. Spon, London.

British Geological Survey, 1986; Geothermal Energy-the potential in the United Kingdom, Downing, R. and Gray, D. (eds), Her Majesty's Stationery Office.

Barbier, E. 1997. Nature and technology of geothermal energy: a review. *Renewable and Sustainable Energy Reviews*, **1**, 1/2, 1–69.

Batchelor, A., Curtis, R., and Ledingham, P. 2005. Country update for the United Kingdom, International Geothermal Association, *World Geothermal Congress 2005*, Turkey.

Bertani, R. 2005. World wide geothermal generation 2000–2005: state of the art, International Geothermal Association, *World Geothermal Congress 2005*, Turkey.

Bertani, R. 2005 World geothermal power generation in the period 2001–2005, *Geothermics*, **34**, 6, 651–90.

Curtis, R., Lund, J., Sanner, B., Rybach, L., and Hellstrom, G. 2005. Ground source heat pumps—geothermal energy for anyone, anywhere: current worldwide activity, International Geothermal Association, *World Geothermal Congress 2005*, Turkey.

Dickson, M. H. and Fanelli, M. 2005. *Geothermal Energy: Utilization and Technology*, Earthscan, London.

DiPippo, R. 2005. *Geothermal PowerPlants: Principles, Applications and Case Studies*, Elsevier, London.

Fridleifsson, G. O. and Elders, W. A. 2005. The Iceland deep drilling project: a search for deep un-conventional geothermal resources. *Geothermics*, **34**, 3.

International Geothermal Association, 2000, Plenary Session III, *World Geothermal Congress 2000*, Kyushu-Tohoku, Japan.

Lund, J., Freeston, D., and Boyd, T. 2005. World wide direct uses of geothermal energy 2005, International Geothermal Association, *World Geothermal Congress 2005*, Turkey.

Lund, J., Freeston D. H., and Boyd, T. L. 2005. Direct application of geothermal energy: 2005 worldwide review, *Geothermics*, **34**, 6, 691–727.

Malolepszy, Z. (ed), 2001; *Geothermal Energy in Underground Mines Conference Proceedings* , University of Silesia, Nov 21–23, 2001, Ustron, Poland.

Mckenna, J. and Blackwell, D. 2005, Geothermal electric power possible from midcontinent and gulf coast hydrocarbon fields, *Oil & Gas Journal*, forthcoming.

Parsons, 1904, President's Address to the Engineering Section, British Association for the Advancement of Science, *Transactions of section G, BAAS Report*, 667–76.

Reed, M., Galwell, K., and Wright, M. 1999. *The Potential for Clean Power from the Earth*, Geothermal Energy Association, USA.

REN21. 2005. *Renewables 2005, Global Status Report*, REN21 Renewable Energy Policy Network, Worldwatch Institute, Washington DC.

Royal Society, 1869; *Report of the Committee for the Purpose of Investigating the Rate of Increase in Underground Temperature*. BAAS Report, Compiled by J. D. Everett, pp. 176–189.

Sigfusson, A. 2003. *Iceland: pioneering the hydrogen economy*, Foreign Service Journal, December, 62–5.

Stefansson, V. 1998. *Estimate of the world geothermal potential*, Geothermal Training Programme, 20th Annual Workshop, Orkustofun, Reykjavik, Iceland.

Stefansson, V. 2000. *Competitive status of geothermal energy in the 21st century*, World Geothermal Congress.

Stosic, N., Kovacevic, A., and Smith, I. 2003. Opportunities for innovation with screw compressors, Proc I Mech E, *Journal of Process Mechanical Engineering*, **217**, 157–70.

Systemes Solaires, 2005. *EU and worldwide geothermal energy inventory*, Systemes Solaires No 170, Barometre Geothermie.

Tester, J. W. 2006, Assessment of the Energy Supply Potential of Engineered Geothermal Systems (EGS)–The Royal Society Discussion Meeting on 'Energy…for the Future', London 2006.

Web Resources

Dickson, M. H. and Fanelli, M. 2004. *What is Geothermal Energy?*, Istituto di Geoscienze e Georisorse, CNR , Pisa, Italy. http://iga.igg.cnr.it/geo/geoenergy.php

International Geothermal Association: http://www.iga.igg.cnr.it/index.php

Geothermal Resources Council: http://www.geothermal.org/index.html

World Geothermal Congress http://www.wgc2005.org

Geothermal Energy Association: http://wwwgeo-energy.org

European Geothermal Energy Council: http://www.geothermie.de/egec_geothermal/menu/frameset.htm

Geothermal Heat Pump Consortium: http://www.geoexchange.org

International Ground Source Heat Pump Association: http://wwwigshpa.okstate.edu

Geo-Heat Center: http://geoheat.oit.edu

The authors

Dr. Tony Batchelor is the Chairman and Managing Director of GeoScience and EarthEnergy Ltd, based in Falmouth. Tony was appointed Managing Director of GeoScience in 1985, with the objective of developing a centre of excellence in skills to support both geothermal engineering and allied work in fields such as oil and gas drilling. He is regularly invited to speak at national and international conferences and as served as Director of the International Geothermal Association; and is the author or co-author of more than sixty papers and three books.

Dr. Robin Curtis is the Technical Manager of EarthEnergy Ltd—part of the GeoScience group. Robin was trained as a nuclear engineer, and has worked on energy from the ground for the past 20 years. Since the early 1990s, he has been instrumental in bringing ground source, or geothermal, heat pumps into vogue in the UK microgeneration sector. As well as being responsible for the design and implementation of a wide range of UK installations, he has generated a number of publications, and has made numerous presentations on the subject.

4. *Wave and tidal power*

Dean L. Millar

Introduction

This chapter reviews how electricity can be generated from waves and tides. The UK is an excellent example, as the British Isles have rich wave and tidal resources.

The technologies for converting wave power into electricity are easily categorized by location type.

1. *Shoreline schemes.* Shoreline Wave Energy Converters (WECs) are installed permanently on shorelines, from where the electricity is easily transmitted and may even meet local demands. They operate most continuously in locations with a low tidal range. A disadvantage is that less power is available compared to nearshore resources because energy is lost as waves reach the shore.

2. *Nearshore schemes.* Nearshore WECs are normally floating structures needing seafloor anchoring or inertial reaction points. The advantages over shoreline WECs are that the energy resource is much larger because nearshore WECs can access long-wavelength waves with greater swell, and the tidal range can be much larger. However, the electricity must be transmitted to the shore, thus raising costs.

3. *Offshore schemes.* Offshore WECs are typically floating structures that usually rely on inertial reaction points. Tidal range effects are insignificant and there is full access to the incident wave energy resource. However, electricity transmission is even more costly.

Tidal power technologies fall into two fundamental categories:

1. *Barrage schemes.* In locations with high tidal range a dam is constructed that creates a basin to impound large volumes of water. Water flows in and out of the basin on flood and ebb tides respectively, passing though high efficiency turbines or sluices or both. The power derives from the potential energy difference in water levels either side of the dam.

2. *Tidal current turbines.* Tidal current turbines (also known as free flow turbines) harness the kinetic energy of water flowing in rivers, estuaries, and oceans. The physical principles are analogous to wind turbines, allowing for the very different density, viscosity, compressibility, and chemistry of water compared to air.

Wave energy resources

Waves are caused by winds, which in the open ocean are often of gale force (speed $>14\,\mathrm{m/s}$). Wave energy resources are thus highest at latitudes between 30° and

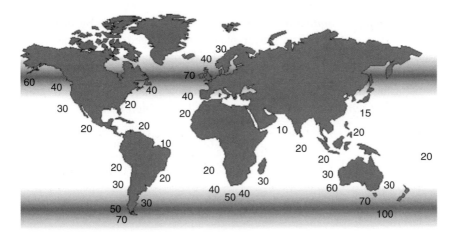

Figure 4.1   World wave power density map (figures show kW/m of wave crest length) highlighting latitudes where resource is highest. Source: CRES, 2002.

60°, north or south, where ocean winds are strongest (Fig. 4.1). Maximum wave strength occurs when the rate of energy transferred from the wind matches the rate at which the wave loses energy from friction, surface current production, and 'whitecapping'. Waves travel great distances in deep water with little energy loss, emerging from storm areas as smooth, regular waves known as 'swell'. The height and length of waves in deep water thus depend on the speed of the wind, and its duration, which is longer for unobstructed expanses of sea (the fetch). Once reaching shallow water (depths less than half the wavelength) waves depend also on the nature of the seabed.

Governments across many countries and regions have commissioned investigations into wave energy technology. In a study by the US Electric Power Research Institute (Previsic *et al.*, 2004) four assessment criteria were used.

1. Design maturity, covering: structural elements, power, mooring, survivability/failure, grid integration, performance, operation and maintenance, deployment, and recovery.

2. Capital cost of purchase of a device.

3. Company viability and willingness to licence the technology.

4. Particular Site Advantage. Wave energy converters exploiting opportunities at specific sites that are not present elsewhere.

Types of wave energy converters

Wave energy converters fall into several classes based on how they interact mechanically with waves. In all cases a reaction force is needed (power production requires *relative* motion) and this is provided either by inertia or by mooring to the seabed. Wave power plants are damaged by extreme weather, so they must be located or moored so as to minimize this happening. For maximum efficiency a WEC must remain 'in phase' with the wave motion. This is achieved with large dimensions and sophisticated motion control systems, each giving cost increases that must be justified by the greater power produced.

Fig. 4.2 shows the different components of wave motion that can be exploited, and Fig. 4.3 shows the different principles of operation of WECs.

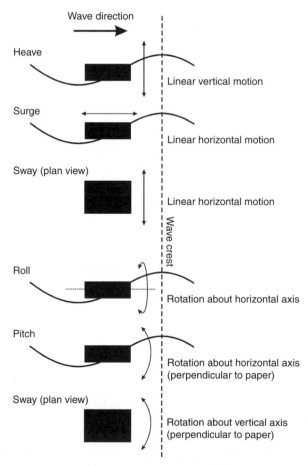

Figure 4.2 The different components of wave motion that can be exploited by wave energy converters.

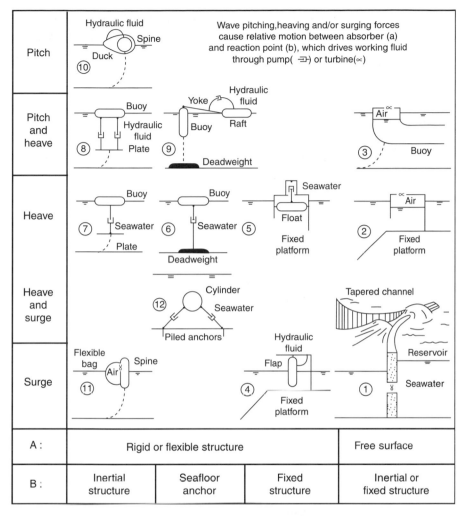

Figure 4.3 Classification of wave energy converters according to A: the way energy is absorbed, B: the type of reaction point. Source: Hagerman, 1995. See also, Falnes, 2003.

Pelamis

The Pelamis, (see Fig. 4.4) developed by Ocean Power Delivery Ltd, UK, most closely resembles concept 9 in Fig. 4.3. The semi-submerged system is 150 m long, weighs 380 tonnes and comprises four 3.5 m diameter cylindrical sections linked by 3 articulating joints. Each joint has a 250 kW rated power module. As waves propagate along the length of the device, the articulating joints actuate hydraulic rams that pump fluid to a smoothing accumulator and on through a hydraulic motor. The hydraulic motor drives a generator to produce electricity. The reaction point for the Pelamis is successive wave crests. The unit is designed to be slack moored in water depths of 50 to 100 m.

Archimedes Wave Swing

The sea-bed-anchored Archimedes Wave Swing (AWS) (type 6 in Fig. 4.3, see also Fig. 4.4) comprises two submerged, floating cylindrical shells, one mounted inside the other (http://www.waveswing.com). As a wave moves through, the outer cylindrical shell moves up and down with respect to the static inner cylinder on a cushion of pressurized trapped air, producing electricity from the motion via a linear generator.

A 2 MW rated AWS was deployed off the Portuguese coast in 2004 and connected to the electricity grid. For tests, the 6 m diameter half-scale prototype was mounted on a static seabed platform. The added mass of the platform produced a total displacement of 7000 tonnes. Rather than a submersible pontoon, commercial devices will utilize a single mooring to anchor the inner cylinder. Thus they will be far less massive.

Group 1 – Development near completion and full-scale long-term testing in the ocean underway

Ocean power delivery—Pelamis

Group 2 – Development near completion, only deployment, recovery and mooring issues are yet to be validated. Construction of full-scale devices is in some cases completed.

Energetech

Wave dragon

Teamwork technology Ltd – Wave swing

Figure 4.4 Examples of wave energy converters in operation or under test, classified according to the US EPRI (Previsic *et al.*, 2004).

WaveDragon

The WaveDragon (see Fig. 4.4) is a sea-bed-anchored, floating WEC that resembles types 1 and 12 of Fig. 4.3 (WaveDragon ApS, 2006). Its parabolic shaped collecting arms face the oncoming wave fronts and focus and channel the waves toward a central ramp, causing the water to enter a reservoir behind it. The water level in the reservoir is thus maintained above the still water level of the open sea. The small head is used to drive water through turbines to produce electricity.

Energetech

The nearshore/shoreline WEC developed by Energetech (see Fig. 4.4) is based on an oscillating water column power take off unit that is located at the focus of a parabolic shaped wave reflector. Entering waves cause the water level to rise rapidly up a confining column, driving air through a turbine. As the water level falls when the wave retreats, air is sucked back through the turbine also producing power on this intake stroke. The parabolic wave reflector amplifies the height of the waves arriving at the column inlet. The complete unit is sea-bed-mounted on a fixed platform, and represented by device 2 in Fig. 4.3.

Energetech prototype units are built from steel, and are 36 m long and 35 m wide with a mass of 485 tonnes. To maximize energy capture they employ a sophisticated control system with a pressure sensor located ahead of the device to tune the device to the period of incident wave phases (Alcorn *et al.*, 2005; *also* http://www.pontediarchimede.com).

Powerbuoy

The Powerbuoy, developed by Ocean Power Technologies (OPT) corresponds most closely to device 5 or 6 in Fig. 4.3. It has an outer cylindrical float that rises and falls under the propagating wave. The power take-off equipment is housed at the bottom of an inner cylinder that maintains reference using anchoring chain and heave reaction plates. The device maximizes energy capture using a 'smart' control system.

The first Powerbuoy prototype was deployed off Atlantic City, New Jersey in 1997, and has evolved through various changes. OPT now has several active projects underway in Hawaii, New Jersey, Spain, France, and the UK (Discovery Channel, 2005).

Comparing wave energy converters—'Wave Hub'

Performances of WECs can be compared both technically and commercially. In absolute terms, WaveDragons have the highest rated capacity, followed by the AWS, then Pelamis. However, when comparing the efficacy of devices their relative scales and fabrication costs need to be taken into account. A good measure of relative efficacy is the annual electrical energy produced divided by device capital cost. When the latter is unavailable, it is possible to use the mass of the device as an analogue for capital cost. Such measures may rank the Pelamis device above the others mentioned, despite it having the lowest rating.

It is necessary to test WEC devices in real situations, involving multiple units operating in array configurations connected to the grid. Choice of a particular WEC depends on how its characteristics are suited to the site resource and how straightforward it is to deploy, moor, recover, and maintain large numbers of identical devices in what are often challenging conditions.

Wave Hub is a development facility for testing marine energy systems. It occupies a 4 km × 2 km area off the north coast of Cornwall, UK, and provides facilities for wave power project developers and WEC device developers to assess the operational performance of arrays of WEC units. The Wave Hub project helps avoid the need for device developers to commit capital for grid connections and planning permission for each WEC technology separately. Also, WEC projects can be financed without too large a 'lone horse' financial risk that might otherwise deter investors. Wave Hub allows comparisons between WEC technologies operating side-by-side in the same resource to determine which technology utilizes the resource most efficiently. Although the global wave energy resource is sufficient to satisfy the world's electricity consumption twice over, exploitable resource sites are relatively scarce at present. Escalation of wave power technology will depend on prioritizing technologies that utilize resources most efficiently and cost effectively.

Tidal energy resources

The nature of tidal energy resources

Tides result from the cyclical nature of the gravitational attraction between the mass of seawater on the Earth's surface and the mass of the Sun and the Moon. We experience tides mainly at shorelines, but even in the absence of land, tides would represent a ∼0.4 m bulge in sea level that circles the globe once every 24 hours and 50 minutes. Tides are therefore akin to waves with a 12.4 hour period: as they approach a shoreline, the reduced water depth causes shoaling, and their height and speed are enhanced as they travel up narrowing estuaries. Headlands and peninsulas can also lead to enhanced tidal current velocities as water is swept around from one tidal basin to another. The world's highest tidal range occurs at the Bay of Fundy, Nova Scotia, Canada (∼18 metres). The predominant type of tide around the world is called semi-diurnal. Locations with semi-diurnal tides have two high-water times and two low-water times per day. Diurnal tide locations have one high-water time and one low-water time each day. Mixed tidal locations experience one tidal cycle with high tidal range and one tidal cycle with low tidal range each day.

Energy recoverable from tidal resources is variable, intermittent, but predictable; the latter characteristic offering an advantage of tidal energy over other forms of renewable energy, with the exception of geothermal energy. The predictable variability arises over various time scales. First and most obviously, the available energy varies over a cycle of approximately 14 days between spring tide and neap tide ranges. Second, the annual energy yield can vary up to 8% from year to year; due to natural variations of the orbital trajectories of celestial bodies driving tides. Third, the recoverable power from a tidal resource varies appreciably within an individual tidal cycle and there will be some occasion during the tidal cycle of 12 hours, 25 minutes—where the output power will fall to zero—and another moment when the output power will peak. These variabilities, although predictable, will not correspond with periodic power demands, and this imposes important constraints on the design of tidal power plants which must incorporate systems for storing the energy or buffering (smoothing out) its magnitude. In addition, tidal height is affected by local weather conditions such as strong winds, which can act with spring tides to produce a tidal surge.

As high-tide times vary along coastlines, one solution is to employ a number of smaller tidal power plants along a coast so that the peak and zero output of each device is staggered in time. This gives, collectively, a more continuous output, albeit with higher costs. This strategy is unavailable to large, barrage-type tidal power plants, but output power can be smoothed by employing multiple adjacent impounded basins. Water flows from one basin to another (and thus output power is maintained) at times in the tidal cycle where a single basin scheme would have zero output power. Double and triple basin barrage schemes are very attractive in this respect and can even overcome some environmental concerns about habitat loss. Multiple basin schemes have the additional benefit of much lower peak output as well as much more continuous operation. Cost-wise, this means that the specifications for switch gear, transformers, and transmission cables are lowered, helping to offset the higher capital requirements for dam construction in comparison to single basin schemes.

Tidal barrage schemes

Tidal barrage schemes impounding huge volumes of water by means of dams, sluices, and locks historically have attracted the most investment attention to date, but this is changing. Like large hydroelectric dams, they introduce significant environmental issues that hinder their development (Kerr, 2005). Table 4.1 lists the world tidal energy resources that are technically accessible for barrage

Table 4.1 Locations suitable for large scale barrage projects.

Country	Region	Mean tidal range (metres)	Approx resource at 'reasonable cost' (GW)	Tide Type
British Isles	Severn/Irish Sea	8	18	Semi-diurnal
France	Normandy/Brittany	7	15	Semi-diurnal
USA	Alaska	7	5	Semi-diurnal Mixed
Canada	Bay of Fundy	10	20	Semi-diurnal
Argentina	Golfo San Matias	5	80	Semi-diurnal
Argentina	Bahia Grande	7	20	Semi-diurnal
Russian Fed.	Beloye More	5	12	Semi-diurnal
India	Gulf of Khambhat	6	8	Semi-diurnal
Australia	Bonteparte Archipelago	6	5	Semi-diurnal
China	Formosa Strait/Hangzhou Wan	5	10	Semi-diurnal
Korea	Korea Bay/Yellow Sea	5	1	Semi-diurnal
Russian Fed.	Sakhalinskiy Zaliv	5	10	Diurnal
Russian Fed.	Penzhinskaya Guba/Zaliv Shelikov	5	30	Diurnal

Sources: Boyle, 2003; Pidwirny, 2005.

projects. The total world installed capacity for tidal power is around 234 GW, and if fully exploited this is estimated to provide 512 TWh/year.

The world's largest operational tidal barrage scheme is the Barrage-de-la-Rance in Brittany, France (mean tidal range: 8.55 m) with an installed capacity of 240 MW. Since 1967, the structure has provided a four-lane, 750 metre-long highway between the towns of St Malo and Dinard. It houses 24 pump-turbine units, sluice gates and lock arrangements for light marine craft, and generates electricity during both ebb and flood tides. The scheme impounds a basin of 22 km^2 and produces 0.64 TWh/annum. The next largest system, rated at 20 MW with output 50 GWh/annum, impounds a 15 km^2 basin at Annapolis Royal, Nova Scotia, Canada where the mean tidal range is 6.4 m. Other schemes in operation are in China and Russia. Total installed tidal barrage generating capacity for China is estimated at 5.7 MW, the main plant being the Jiangxia Tidal Power Station in Leqing Bay (tidal range: 5.1 m, 3.9 MW, 5 GWh/annum). In Russia, the 400 kW Kisalaya Guba power plant is located north of Murmansk in a gulf opening to the Barents Sea.

Although the number of tidal barrage schemes actually operational and feeding electricity into a grid are still few in number, there is great activity in feasibility studies and project proposals. The range is huge, with the largest being a 87.4 GW proposal that would impound a sea area of 20,500 km^2 in Penzhinskaya Bay, Siberia.

Environmental impacts

The La Rance scheme was completed in 1967. During its construction the river was isolated from the sea for approximately 3 years, and this led to the death of many species of plants and animals that had previously flourished in estuarine and mudflat environments. Populations started to recover once the connection with the sea was re-established, and more so when the barrage started operations although modifications were needed to alleviate silting. The barrage also exacerbated the problem of algal blooms caused by nitrate based fertilisers. Prior to construction of the barrage, La Rance was an important route for the transport of goods inland from the port of St Malo. Towns along the river and canal systems that depended on this trade have had to adapt and goods that were originally transported along the river are now displaced to the roads. Despite these negative aspects, a tour of the La Rance basin today reveals a very attractive environment that appeals to tourists and regular leisure water users. Habitats and floral and faunal populations have recovered.

A study of the environmental impacts of a tidal barrage power station constructed across the Severn Estuary (UK), concluded that a barrage may actually be beneficial because it reduces the speed of tidal currents on the river/sea bed to a level at which it can be colonized by species currently unable to obtain an anchoring (Kirby and Shaw, 2005). Some studies, for example by Dadswell and Rulifson, 1994, suggest that fish and other marine animals may be detrimentally affected by turbine motion in barrage schemes. The general consensus is to examine alternative methods of harnessing tidal energy. Thus tidal current turbine technology that operates in the free stream of tidal flows without the need for impounding dams has attracted significant interest and investment in recent years.

Tidal current turbines

In contrast to barrage projects which harness the potential energy of a large head of water, tidal current turbines harness the kinetic energy of freely flowing water. The kinetic energy resource is similar in magnitude to the potential energy resource harnessed in a tidal barrage scheme. The locations where enhanced concentrations of kinetic energy are available broadly coincide with identified locations for barrage projects. However, tidal flow round coastline obstructions, such as headlands and peninsulas, enhance current speed and provide locations where tidal current turbine technologies can be exploited whereas tidal barrage technology cannot. The threshold peak tidal current velocity for an exploitable resource is about 1.5 m/s.

The largest rating of a tidal current turbine currently being considered is 1 MW, rather less than that of the two largest operating barrage schemes. A large increase in scale is thus required: for example, had tidal current turbines been installed in

La Rance estuary, 240 of the largest units would have to be installed to reach the same installed capacity and similar annual electrical power yield.

There are several advantages of tidal current turbine developments over tidal barrage projects.

1. As more turbine units are required, there are greater opportunities for economies of scale in turbine manufacturing and lowered capital costs than with tidal barrage systems.

2. Tidal current turbines are more flexible in respect of the locations where they can be deployed.

3. Tidal current turbine farms do not require water-impounding structures and are thus less restricted by environmental impact problems.

4. Removing the requirement for dam construction means that tidal current turbine developments can be smaller and less capital intensive (but not necessarily cheaper, in terms of £/kW installed capacity).

The physics of tidal current turbine technology is similar to that for wind turbines, except that the density and viscosity of water are much higher than air. Because of this similarity, and because institutional investors understand and are already convinced about wind power developments, there is a close financial and commercial development analogue for developers of tidal current turbines.

Turbine technologies under development

Classification of conventional tidal current turbines types closely mirrors those for wind turbines. The main distinction is whether the rotor axis of the turbine is mounted horizontally or vertically. Alternatively, the energy in tidal currents can be extracted by hydroplanes or harnessed indirectly, by devices (venturi) that convert water flow into air flow, which is then used to drive the generator.

Horizontal axis turbines

The choice of rotor size, and hence the power rating, depends primarily on the depth of water in which the turbine is to be deployed (Bryden *et al.*, 1998). Provided shipping can be excluded from the vicinity of the turbine installation then the rotor axis depth and the rotor diameter can be half the lowest depth of water at low tide.

The world's first grid-connected horizontal axis tidal current turbine was installed in Norway in 2003 to supply Hammerfest, the world's most northerly town. The 300 kW rated turbine has a 22 m diameter three-blade rotor and is mounted on a 120 tonne steel tripod structure that rests on the bed of the Kvalsundet strait and is stabilized by additional weights totalling 200 tonnes. The 54 tonne nascelle contains the generator, gearbox, drive train, and control components, including a pitch regulating mechanism that allows the unit to harness reversing tidal currents. Within the strait, which is 400 m wide at its narrowest point and 50 m deep, the average tidal current velocity is 1.8 m/s (peak of 2.5 m/s) and each unit can provide 0.7 GWh of electricity per year.

SeaFlow and SeaGen   SeaFlow is a prototype horizontal axis tidal current turbine that has an 11 m diameter, two-blade rotor with variable pitch. It is rated at 300 kW, attainable when the tidal current is 2.7 m/s or greater. The rotor, drive train, and generator are mounted on a chassis that can be raised out of the water to allow inspection and maintenance without recourse to dive teams. The chassis travels up and down a vertical steel pile drilled into the seabed. The prototype was designed to consider only a single tidal current direction in testing. The power produced in tests has been reported to be 27% greater than predicted, and this success has increased confidence in tidal current turbine technology and secured further funding from the UK government for the developers, Marine Current Turbines Ltd. The next stage is a device called SeaGen which has twin rotors, each rated at 500 kW, which operate when the tide flows in either direction.

A turbine farm comprised of multiple 1 MW SeaGen units is under consideration, as depicted in Fig. 4.5. The rating per unit is unlikely to increase much above 1 MW: SeaGen deployment depths are limited to 40 m, so that the rotor size cannot be greater than 20 m in areas where marine traffic will be excluded, and around 15 m where marine traffic will retain access. A project with 133 MW capacity would comprise 133 SeaGen units, spaced 60 m apart to form a linear array 8 km long, and could provide an output of 466 GWh/annum.

Vertical axis turbines

One important advantage of vertical axis tidal current turbines is that they require no blade pitch control to allow for reversing of the tidal current direction; the rotor rotates the same way irrespective of the direction of the impinging fluid.

Davis turbines   In the free stream Davis hydroturbine, developed since 1981 by Blue Energy Inc. of Canada, the axis of rotation is vertical and the rotor comprises 3 or 4 vertical, straight hydrofoil-profile blades mounted on radial arms along the length of the runner. Davis turbines operate in either ducted or unducted conditions as stand-alone units or as units in a 'tidal fence'. In many cases, the gearbox and generator can be located on top of the rotor, out of the water, which simplifies routine maintenance. To date, six prototypes have been

Figure 4.5 Artist's impression of a tidal current turbine farm using horizontal axis turbines (Source: www.marineturbines.com).

tested, ranging from small laboratory scale test units to a 20 kW unit connected to the Niagra Power Corporation grid in 1983 and a 100 kW unit connected to the Nova Scotia Power Corporation grid in 1984 (Takenouchi *et al.*, 2006).

Blue Energy Inc. of Canada have been involved with various tidal energy development projects, including a 500 kW demonstration plant of floating Davis turbine units off British Columbia, Canada and a 4 km long, 274 turbine, 2200 MW rating tidal fence concept that would cross the Strait of San Bernardino in the Philippines.

Gorlov Helical Turbine The Gorlov turbine (Gorlov, 2001; Gorban *et al.*, 2001) is a cross flow turbine with helical blades that run around a cylindrical surface with each blade having a hydrofoil profile. It shares the key advantage of the Davis turbine over many of the horizontal axis turbine units that, due to its axial symmetry, there is no need for blade pitch control. However, vertical axis hydro turbine units can become unstable at high speed, requiring shut down when the resource is highest. This problem can be overcome in Gorlov turbines because the helical blade geometry produces more stable running.

Gorlov turbines feature in a large 90 MW project planned in Korea: a tidal fence development bridging the Uldolmok Strait between the Korean mainland and Jindo Island where peak tidal currents are 6.5 m/s.

Cyclodial Turbines   The Kobold turbine is a three bladed vertical axis device produced by Enermar. The pitch of the blades is variable and automatically regulated though means of the acting centrifugal forces and an asymmetrical blade profile. The rotor transmits power through a very high ratio (1:160) gearbox to a permanent magnet generator. The Enermar system is being applied in the Strait of Jintang (Zhoushan Archipelago) in China.

Hydrofoil Devices

Stingray   The Stingray (http://www.engb.com/) uses a variable pitch hydroplane to extract energy from tidal currents. The hydroplane is attached to a support arm that rises and falls according to the sense of the hydroplane lift. This produces a 'pumping' motion to actuate hydraulic cylinders that pressurize oil and turn a hydraulic motor that runs the generator. The pitch of the hydroplane is adjusted continuously through each stroke of the supporting arm such that at the top of its stroke, the lift force acts downward. At the bottom of its stroke, the hydroplane pitch is again adjusted so that the lift force acts upward.

SeaSnail   The SeaSnail is designed primarily as a seabed platform for tidal current turbine devices. The platform is an evolution of the Stingray device in that it uses hydroplanes to create down-force to 'pin' the platform structure to the seabed. The enhanced down-force creates a greater resistance to shear loads that are primarily introduced as reaction thrusts from the energy capture components. The device is 15 m long by 12 m wide and has a mass of 30 tonnes. The six hydroplanes mounted on the chassis create a force equivalent to 200 tonnes. Although the SeaSnail prototype features a simple horizontal axis tidal current turbine for demonstration purposes, it is a move away from simply relying on steel or concrete mass to anchor seabed mounted structures, and will allow such marine energy capture units to be fabricated more cheaply.

Venturi Devices

Hydroventuri   A device from Hydroventuri Ltd utilizes the static pressure drop occurring at the throat of a venturi-shaped duct which draws in air as water flows through. This air passes through a turbine, the shaft of which is connected to a generator. The advantage over other tidal current turbines is that the submergible components have no moving parts; the turbine and generator do not need to be mounted on a surface platform. The disadvantage is that a connection to atmospheric air to the submerged components needs to be maintained.

Table 4.2 Ranking of wave and tidal devices based on estimates of their G-value (*annual yield*/capital cost ratio).

Device	annual kWh/tonne device mass)
(1) SeaGen	9281
(2) SeaFlow	8002
(3) Stingray on SeaSnail	6488
(4) Pelamis (WEC)	4034
(5) Rotech 1.5	3694
(6) Stingray	2884
(7) Powerbuoy (WEC)	2061
(8) Archimedes WaveSwing (WEC)	1184
(9) WaveDragon (WEC)	521

Ranking of marine renewable energy technologies

Wave and tidal current devices can be ranked on the basis of their G-values, the ratio of the annual yield to the mass of the device (mass being an analogue for capital cost). An example of such a ranking is given in Table 4.2.

Despite its simplicity (the ranking assumes that cost relates directly to mass, and there are uncertainties in estimating the mass of some of the devices) it allows comparisons across many power conversion concepts and different forms of marine renewable energy. Current competition is strong for marine renewables, driven by commercial interests. Ultimately, four key factors will establish the system of choice for exploiting marine energy resources:

- Accessibility of the resource.

- Simplicity of power conversion concept—the WaveDragon is easily the best in this respect.

- Ease of operations and maintenance—devices running without diver intervention are best.

- Proven survivability and efficacy of mooring arrangements—devices that break free from moorings and cause serious marine accidents may be banned from the sea.

Assessment of wave and tidal current resources (UK)

Based on the Wave Hub model, the criteria for suitable sites for wave farms are easily identifiable:

- Average annual wave power density $>25\,\text{kW/m}$.

- Water depth of 50–60 m.

- Proximity of electricity distribution network with available capacity.

- Outside major shipping routes.

- Wave energy arrives at the site predominantly from one direction.

- Proximity to onshore facilities to support operations and maintenance activities.

The site criteria for tidal turbine farms (based on SeaGen) are likewise summarized as follows:

- Peak current velocity on mean spring tide 2.5–3.0 m/s, in a normal current shear location.

- Water depth 30–35 metres at Lowest Astronomical Tide (LAT).

- Proximity of electricity distribution network with available capacity.

- Outside major shipping routes.

- Predominantly bi-directional tidal current directions—narrow tide ellipses.

- Relatively low ratio of peak spring current speeds to peak neap current speeds.

- Proximity to onshore facilities to support operations and maintenance activities.

Figs. 4.6 and 4.7 illustrate criteria applicable to the respective technologies and an estimate of sites allowing either a Wave Hub wave farm model or a large scale tidal current turbine farm to be considered. The numbers of sites identified against each model are summarized in Table 4.3. The tidal current resource is clearly the more significant both in terms of installed capacity and annual output. The reasons for this are threefold: there are more resource areas, the installed capacity per farm is over 100 MW higher, and the capacity factors of the devices are higher.

The simplistic analysis illustrates a crucial factor restricting growth of either sector—access to an electricity transmission network. For the Hebridean and Orkney Islands, where both tidal current and wave power opportunities are appreciable, development of either type would justify extending the grid.

This accessibility has been noticed by developers and regional public sector organizations and led to the Wave Hub project in Cornwall and the proposed WaveDragon deployment off Milford Haven. These regions have the resource and grid capacity to support one more development each. However, beyond this, the wave sector at least may find it very difficult to locate suitable development sites without major transmission and distribution network upgrades.

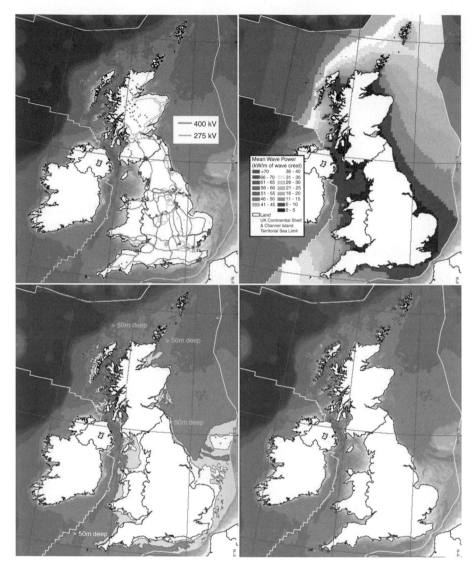

Figure 4.6   Wave energy resource assessment coverages of the UK. Upper left: UK electricity transmission grid. Upper right: Mean wave power density. Lower left: Boundaries of outer shaded areas identify the 50 m water depth contour. Lower right: Sites identified as suitable for wave power developments. Source for all base maps: ABPMer *et al.*, 2004.

The situation is less severe for tidal current turbine developments because the north Somerset/north Devon, south Wales and Hampshire/Dorset coasts are well served with grid infrastructure where the resource is at its highest.

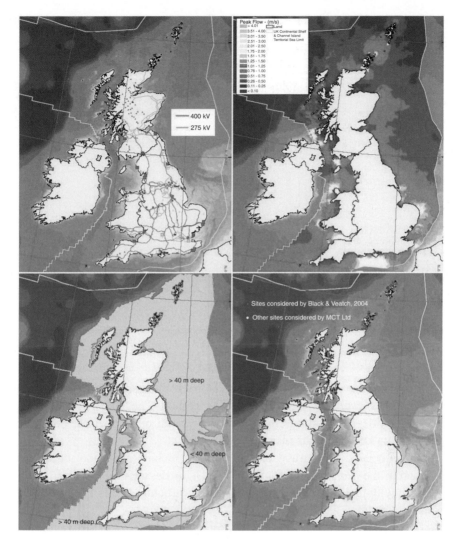

Figure 4.7 Tidal current resource assessment coverages of the UK. Upper left: UK electricity transmission grid. Upper right: Peak current speed on a spring tide. Lower left: Light grey areas show water depths >40 m, areas further inshore are <40 m deep. Lower right: Sites identified by Black & Veatch, 2004 and MCT Ltd as suitable for tidal current turbine arrays. Source for all base maps: ABPMer *et al.*, 2004.

Summary and conclusions

This article has reviewed tide and wave energy resources and technologies that exploit them. For each resource, an interpretation of how these technologies could be deployed at large scale in UK waters followed.

Table 4.3 Summary of simplified UK resource assessment of wave and tidal current energy.

Region	Number of tidal current sites		Number of wave sites	
	With present grid	With grid upgrade	With present grid	With grid upgrade
Shetlands	0	4	0	6
Orkney	11	11	4	4
Hebrides	1	7	0	19
Islay/Kintyre	1	6	0	1
N Ireland	4	4	0	0
Wales	3	4	2	2
SW England	4	9	2	4
E England	1	2	0	0
Total number of sites	25	50	8	36
Installed capacity per farm (MW)	133	133	30	30
Annual output per farm (GWh)	466	466	53	53
Total installed capacity (MW)	3325	6650	240	1080
Total annual output (GWh)	11.7	23.3	0.4	1.9

Pelamis is the leading WEC based on annual kWh produced/mass of device, although the Powerbuoy device may rank higher. The North Atlantic rated WaveDragon device produces more than any of the other devices, but suffers in the ranking because of its mass (33,000 tonnes).

Costs associated with wave or tidal current technologies may be lowered by adopting the SeaSnail platform concept. This device replaces dead weight anchoring with hydroplane down-force, using a structure that is a fraction of the corresponding dead weight mass.

In the UK, the accessible tidal energy resource is greater than the wave energy resource; although worldwide the wave resource is greater. With the current grid infrastructure, tidal current turbine farms with a total installed capacity of 3.3 GW may be foreseen, contributing around 11.7 TWh of electric power. Under the same conditions, the capacity of WECs could reach 240 MW, contributing around 0.4 TWh per annum of electric power.

For both technologies, an important future barrier to exploitation is the proximity and accessibility of the transmission and distribution grid. Without a grid upgrade (particularly in Scotland) suitable sites may become rather scarce. Under this scenario, which would lead to areas such as Shetland, the Hebridean Islands, and some parts of south west Scotland being included in the accessible resource, installed capacity could total 6.6 GW for tidal current farms and 1.1 GW for wave farm developments. Annual yield would stand at 23.3 TWh for tidal power and 1.9 TWh for wave power respectively.

The UK has set up a chain of useful facilities to support marine energy developments. The New and Renewable Energy Centre (NaREC) allows testing of physical models at small scale; the European Marine Energy Centre (EMEC) in Orkney provides facilities to test individual full scale prototypes; and the Wave Hub provides facilities for pre-commercial testing of arrays of wave power conversion devices. For the UK, the marine renewables sector is being driven by financial incentives. A similar situation exists in Portugal and to a lesser extent in the US where individual states offer specific incentives. A great deal of commercially driven development activity is thus emerging.

Resources and further information

ABPMer, The Met Office, Garrad Hassan and Proundman Oceanic Laboratory, 2004. *Atlas of UK Marine Renewable Energy Resources*: Technical Report, A strategic environmental assessment report. Prepared for the UK Department of Trade and Industry, Project Ref No. R/3387/5, Report No. 1106.

Alcorn, R., Hunter, S., Signorelli, C., Obeyesekera, R., Finnigan, T., and Denniss, T., 2005. *Results of the Testing of the Energetech Wave Energy Plant at Port Kembla on October 26, 2005*. Available at http://www.energetech.com.au/attachments/Results_PK_Wave_Energy_Trial.pdf.

Boyle, G. (ed), 2003. *Renewable Energy, Power for a Sustainable Future*, 2*nd* edition. Oxford University Press, Oxford.

Bryden, I., Naik, S., Fraenkel, P., and Bullen, C.R., 1998. Matching tidal current plants to local flow conditions. *Energy 23*, 9, 699-709.

Centre for Renewable Energy Sources (CRES), 2002. *Wave Energy Utilisation in Europe, Current Status and Perspectives*. Report from the European Thematic Network on Wave Energy. Available online at: http://www.cres.gr/kape/publications/download_uk.htm.

Dadswell, M. J., and Rulifson, R. A., 1994. Macrotidal estuaries: a region of collision between migratory marine animals and tidal power development. *Bio. J. Linnean Soc. of London*, **51**, 1–2, 93–113.

Discovery Channel, 2005, Video article on OPT Powerbuoy developments. [Online] http://www.exn.ca/video/?video=exn20051114-buoy.asx.

Engineering Business Ltd, 2004. *Project Summary: Stingray Tidal Stream Energy Device–Phase 3*. Report number: M04-102-01. Available at: http://www.engb.com/Downloads/papers/.

Falnes, J., 2003. *Principles for Capture of Energy from Ocean Waves*. Department of Physics, Norges Teknisk Naturvitenskapelige Universitet (NTNU) (Norwegian University of Science & Technology) http://www.phys.ntnu.no/instdef/prosjekter/bolgeenergi/phcontrl.pdf. Accessed 10th April 2003.

Gorban, A. N., Gorlov, A. M., Silantyev, V. M., 2001. Limits of the Turbine Efficiency for Free Fluid Flow. *J. Energy Re. Tech.* **123**, 311–17.

Gorlov, 2001. *Tidal Energy.* Conference paper available from http://www.gcktechnology.com/GCK/Images/ms0032%20final.pdf.

Hagerman, G. 1995. A standard economic assessment methodology for renewable ocean energy projects. *Proceedings of the International Symposium on Coastal Ocean Space Utilization,* COSU, 129–38.

Kerr, D. 2005. Proceedings of Institute of Civil Engineers, *Civil Engineering* **158**, 32–9.

Kirby R., and Shaw, T. L., 2005. Severn Barrage, UK—environmental reappraisal. *Proc. Inst. Civ. Eng., Engineering Sustainability,* **158**, 31–9.

Kofoed, J. P., Frigaard, P., Friis-Madsen, E., and Sørensen, H. C., 2004. Prototype Testing of the Wave Energy Converter Wave Dragon. *Proc. 8th World Renewable Energy Cong.,* A. A. M Sayigh (ed). Denver, Colorado, Aug 28th- Sept 3rd 2004.

Pidwirny, 2005. *Fundamentals of Physical Geography,* Chapter 8, Ocean tides. Online book available at: http://www.physicalgeography.net/fundamentals/contents.html Ponti di Archimede SpA, 2006. http://www.pontediarchimede.com/.

Previsic, M., Bedard, R., and Hagerman, G., 2004. *E21 EPRI Assessment, Offshore Wave Energy Devices.* US Electrical Power Research Institute Report No. E21-EPRI-WP-004-US-Rev1. Available online at: www.epri.com. [Accessed 09 Apr 2006]

Takenouchi, K., Okuma, K., Furukawa, A., and Setoguchi, T., 2006. On applicability of reciprocating flow turbines developed for wave power to tidal power conversion. *Renewable Energy,* **31**, 209–23.

Teamwork Technology BV, 2004. http://www.waveswing.com/ [Accessed 25 May 2004] (Archimedes Wave Swing device)

TideElectric, 2005. http://www.tidalelectric.com/Tidal%20Resource.htm

WaveDragon ApS, 2006. http://www.wavedragon.net/.

The author

Dean L. Millar is a senior lecturer at the Camborne School of Mines, University of Exeter. Originally trained as a mining engineer, practising in the UK and overseas, he specialized in geotechnical engineering and rock mechanics, lecturing in this subject at Imperial College London between 1993 and 1997. An interest in the coastal engineering applications of rock mechanics led to his appointment at the Camborne School of Mines in Cornwall, where he has conducted research into the use of abandoned mine shafts as oscillating water column wave energy conversion devices and wave energy resource assessment and impact. He maintains a broad interest in all aspects of renewable energy in addition to wave energy, particularly transport biofuels, and established the UK's first undergraduate degree programme in renewable energy in 2003.

5. *Wind energy*

Bill Leithead

Introduction

A wind turbine or even a wind farm, i.e. a group of wind turbines, is becoming an increasingly familiar sight in the countryside today. The wind turbine converts the power in the wind to electrical power and consists of a tower, rotor, typically with three blades as in Fig. 5.1, and a nacelle containing the power converter.

From its rebirth in the early 1980s, wind power has experienced a dramatic development. Today, other than hydropower, it is the most important of the renewable sources of power. With an installed capacity equivalent to that required to provide electricity for over 19,000,000 average European homes and annual turnover greater than £5,500,000,000, wind energy has exceeded its year-on-year targets over the last decade. This growth in the contribution to electricity generation from wind power in Europe is likely to continue over the next few years, since the EU Commission has set a European target for 2010 of 12% of electricity generation from renewable sources. In the long term, the achievable limit to the contribution of wind power is estimated to be 30% of the total European demand, an amount almost equal to the installed nuclear capacity.

In the UK, wind power is the fastest growing energy sector. Over 4,000 people are employed by companies working in the wind sector, and it is estimated by the UK Department of Trade and Industry (DTI) that the next round of offshore wind development could generate a further 20,000 jobs. In a 2003 Energy White Paper, the UK government aspired to achieving a 60% reduction in UK CO_2 emissions

Figure 5.1   A typical modern wind turbine.

by 2050. In order to do so, it has set targets for UK electricity generation from renewable sources of 10% of electricity demand by 2010 and 20% by 2015. Since it is the most mature of the renewable energies, much of these near term targets must be met by wind power. Irrespective of whether these targets are achieved, the potential for increase in the UK is substantial.

The prospects for wind power development in the UK are dependent on the available wind resource, public acceptance, and technical development. Each of these issues is discussed below.

Wind energy resource

The power in the wind increases with the cube of the wind speed. Hence, the resource available at a specific location depends strongly on the annual mean wind speed. The European Wind Atlas, Fig. 5.2, indicates on a large scale those regions with similar annual mean wind speed. However, the annual mean wind speed at a specific location depends strongly on the local topography and may vary considerably from the regional mean. It should also be noted that the annual

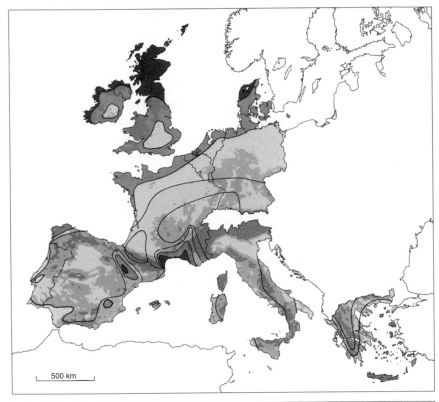

Wind resources at 50 metres above ground level for five different topographic conditions										
	Sheltered terrain		Open plain		At a sea coast		Open sea		Hills and ridges	
	ms⁻¹	Wm⁻²	ms⁻¹	Wm⁻²	ms⁻¹	Wm⁻²	ms⁻¹	Wm⁻²	ms⁻¹	Wm⁻²
	> 6.0	> 250	> 7.5	> 500	> 8.5	> 700	> 9.0	> 800	> 11.5	> 1800
	5.0-6.0	150-250	6.5-7.5	300-500	7.0-8.5	400-700	8.0-9.0	600-800	10.0-11.5	1200-1800
	4.5-5.0	100-150	5.5-6.5	200-300	6.0-7.0	250-400	7.0-8.0	400-600	8.5-10.0	700-1200
	3.5-4.5	50-100	4.5-5.5	100-200	5.0-6.0	150-250	5.5-7.0	200-400	7.0-8.5	400-700
	< 3.5	< 50	< 4.5	< 100	< 5.0	< 150	< 5.5	< 200	< 7.0	< 400

Figure 5.2 European on-shore wind atlas.

mean wind speed increases with the height above ground of the measurement; for example, the difference in annual mean wind speed between a height of 40 m and 60 m is approximately 10%. Hence, the resource at a specific site depends strongly on the locality and the size of wind turbines to be installed. Nevertheless, general trends hold, with the annual mean wind speed increasing to the north and to the west.

With an annual mean wind speed of roughly 7 m/s for England and Wales and 8 m/s for Scotland, the UK clearly has a very rich resource compared to other European countries. To emphasize this fact, it is sometimes claimed that the UK has 40 % of the European wind resource. Of course, this claim is highly subjective, being dependent on the details of the resource estimation and the definition of Europe used, and should be interpreted merely as indicative of the comparative richness of the UK resource. Not all of the wind power resource can be realized however. Wind farms must be sited in open country avoiding sites of scientific interest and sufficiently far from inhabited buildings and roads. Unfortunately, much of the exploitable resource is located to the north and west, far from the main population centres. To fully exploit this resource would require the transmission of large amounts of electrical power over long distances from the points of generation to the points of consumption. The national grid is currently not well placed to accommodate this transmission because of such restrictions as the capacity of the inter-connector between Scotland and England.

The European Offshore Wind Atlas, Fig. 5.3, has a similar pattern to Fig. 5.2. Again, the wind power annual mean wind speed increases to the north and west, with the UK well resourced. The resource that can be exploited must be in shallow water and away from major shipping lanes. Even with these restrictions, there is a large exploitable resource off the north west, south east and south west coasts of England. These have the advantage of not being far away from the large centres of population, and so may be particularly valuable. In particular, there is a considerable offshore resource round the southern part of England. For example, from Fig. 5.2 and 5.3, the annual mean wind speed at a height of 50 m on the sea coast round England is 7–8.5 m/s and 10 km offshore at a height of 100 m is 8.5–10 m/s. Sites with these wind speed attributes would be sufficiently well resourced for exploitation.

In comparison to the UK's annual electricity demand of roughly 350 TWhours/year, it would be technically feasible, but not practical, to generate 1,000 TWhours/year of electricity from wind. Instead, the accessible and economic resource is approximately 150 TWhours/year. Onshore wind power could contribute in the region of 50 Twhours/year and offshore wind power could contribute in the medium term 100 Twhours/year.

From the above discussion, it is clear that the UK wind power resource is particularly strong in comparison to other European countries. However, the UK record in exploiting that resource is relatively poor. The very rapid growth in wind power capacity installed in the EU is shown in Fig. 5.4. By 2005, it stood at 35 GW, constituting 70% of the world total. The USA accounts for much of the non-European capacity. The contributions to the total EU installed capacity by the leading countries, Germany, Spain, and Denmark, are 16,629 MW, 8,263 MW, and 3,117 MW, respectively, followed by Italy, the Netherlands, and the UK, with 1,125 MW, 1,078 MW, and 888 MW. Although it is increasing rapidly, the UK installed capacity is rather modest in comparison to Germany, Spain, and Denmark, especially when the extent of the resource is taken into account

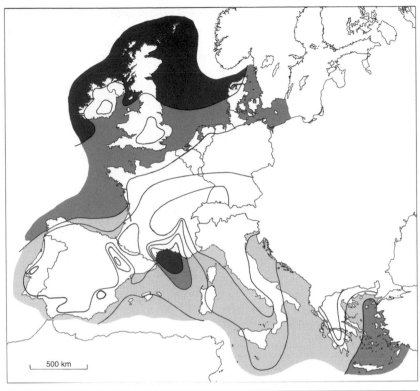

Wind resources over open sea (more than 10 km offshore) for five standard heights									
10 m		25 m		50 m		100 m		200 m	
ms^{-1}	Wm^{-2}	ms^{-1}	Wm^{-2}	ms^{-1}	Wm^{-2}	ms^{-1}	Wm^{-2}	ms^{-1}	Wm^{-2}
> 8.0	> 600	> 8.5	> 700	> 9.0	> 800	> 10.0	> 1100	> 11.0	> 1500
7.0-8.0	350-600	7.5-8.5	450-700	8.0-9.0	600-800	8.5-10.0	650-1100	9.5-11.0	900-1500
6.0-7.0	250-300	6.5-7.5	300-450	7.0-8.0	400-600	7.5-8.5	450-650	8.0-9.5	600-900
4.5-6.0	100-250	5.0-6.5	150-300	5.5-7.0	200-400	6.0-7.5	250-450	6.5-8.0	300-600
< 4.5	< 100	< 5.0	< 150	< 5.5	< 200	< 6.0	< 250	< 6.5	< 300

Figure 5.3 European offshore wind atlas.

(see Figs. 5.2 and 5.3). To highlight this, the contribution of wind-generated electricity to the total annual consumption for several EU countries is listed in Table 5.1., where the absolute values are given together with the wind generation as a percentage of the annual consumption. In terms of the latter, only in Denmark, Spain, and Germany does wind power contribute significantly towards electricity supply. In addition, the fraction of the potential, i.e. the accessible and economic resource, exploited to date, is listed in Table 5.1. The UK is only exploiting 1% of the usable resource.

It might be expected that the explosive growth of wind power indicated by Fig. 5.4 would be accompanied by a reduction in the cost. The price of wind turbines in €/kW is also shown in Fig. 5.4 and confirms the expected

Table 5.1 Wind generation for several EU countries.

	Annual Consumption (TWh)	Wind Generation (TWh)	Wind Generation %	Fraction of Potential %
Austria	60.15	0.24	0.4	8
Denmark	81.73	5.28	6	18
France	431.86	0.20	0.04	0.2
Germany	531.78	18.49	3.47	77
Spain	221.42	11.95	5	14
UK	349.20	1.45	0.4	1
Total	2562.7	42.60	1	6.6

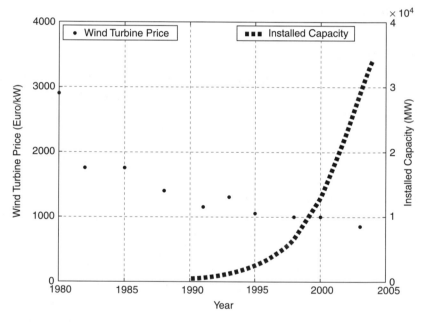

Figure 5.4   Growth of installed capacity and reduction in price.

reduction. It is due in part to improvements in the technology but mainly to economy of scale derived from large-scale production. As a result, the price of electricity generated by wind power is now becoming competitive. Although the price is development specific, it is approximately 4.5–6.0 c€/kW hour for onshore wind farms. The price for offshore wind farms is estimated to be 50% higher. For comparison, the price of electricity for new coal generation plant is approximately 3.8–6.8 c€/kWhour and, for a new nuclear generation plant, 6.0–10.5 c€/kWhour. Associated with the generation of electricity are external costs that are not accounted for in the price figures quoted above. These include the cost of environmental impact and decommissioning of the plant. The external costs for wind power are equivalent to 0.26 c€/kWhour. The corresponding figure for conventional coal generation, although heavily dependent on the assumptions made concerning the environmental costs related to carbon emission, is 2–15 c€/kWhour. Furthermore, the scrap value of the wind turbines pays for decommissioning, and the energy payback time— i.e. the time taken to generate the quantity of power used during the manufacture and installation of a wind turbine—is only 3 to 10 months. In other words, over a lifetime of 20 years, the power produced is at least 24 times the power consumed during its manufacturing and installation. Wind power is already almost cost competitive with conventional sources of electricity generation and will become increasingly so as the latter become more expensive in the future.

Public acceptance

Whenever a wind farm development is proposed, concerns arise over its impact on the local environment. Three issues that are usually raised are noise, the effect on birds, and visual intrusion in the landscape. These are not entirely amenable to a technical solution, as perception of them is to some extent subjective. Public acceptance of wind power depends on a sensitive response by the industry.

Noise emitted by wind turbines has two main sources, mechanical noise and aerodynamic noise. The source of mechanical noise is largely the gearbox. It may be amplified through resonating with the tower. However, in a well-designed wind turbine, it can be reduced to an unobtrusive level through improved damping and insulation of the gearbox and modification of the resonance characteristics of the tower. The aerodynamic noise is at low frequency and is dependent on the rotational speed of the wind turbine rotor. In high wind speeds, it is masked by the ambient noise but, at low wind speeds, it is more apparent. Some reduction in aerodynamic noise has been achieved through improvement to the aerodynamic design of the wind turbines and through operating the machines with lower rotational speeds in low wind speed. The latter strategy is being adopted on most modern wind turbines. The measured noise level of a 1 MW wind turbine at 300 m is 45 dB. It is less noisy than a vacuum cleaner at 30 m, which is measured at 50 dB, in other words, 'half as noisy'. However, the characteristics of a noise are also important to its perception, and wind turbine noise is perceived by some people to be more intrusive than other sources. Indeed, the awareness of wind turbine noise varies greatly and appears to be partly psychological.

There are concerns that wind farm developments will result in the deaths of many birds from collisions with the machines. Data for bird strikes are available from countries with an established wind industry. During 2003, there were 88 deaths of medium and large birds caused by 18 wind farms in Navarra, Spain; that is, the annual mortality rate is 0.13 birds/turbine or, on average, a wind turbine kills a bird every seven years. In Finland with 82 MW of installed capacity, during 2002, there were 10 bird fatalities from collision with wind turbines, compared to 820,000 birds killed annually from collision with other artificial structures (cars, buildings, etc.). In the USA, during 2002, there were 33,000 bird fatalities due to wind turbines, but 100 million to 1,000 million from collisions with artificial structures. In the UK, domestic cats are thought to kill 55 million birds annually. Further evidence in support of these low fatality rates is obtained from visual observation and radar observation studies of bird flight. It is observed that, when flying through a wind farm, birds tend to avoid the turbines, keeping as far away from them as possible. Nevertheless, concern over the impact of wind farms on birds remains, not necessarily due to the fatalities, but due to habitat loss through displacement of the birds.

A common perception is that, for wind power to make a substantial contribution to the UK energy needs, large numbers of wind turbines occupying an extensive part of the countryside are required. The result would be a great loss of amenity due to visual intrusion in the landscape. However, the number of machines involved is frequently exaggerated. Just 600 modern 5 MW wind turbines would be sufficient to replace 1,000 MW of conventional generation; that is, to supply 20% of the peak Scottish electricity demand. To locate sufficient wind farms to provide 10% of the UK's electricity needs would require a land area of 30 km by 40 km, less than 0.5% of the land area of the UK. Furthermore, the land used for a wind farm is not excluded from other use, e.g. agricultural land would remain so except for a small footprint round the base of each turbine and the access roads to the wind farm. Hence, even with large-scale development, the land that would be occupied by wind farms is not overly extensive, particularly since a substantial part is likely to be sited offshore and, thereby, rendered less visually intrusive. The visual impact of offshore wind farms quickly diminishes with distance, and 10 km would suffice. Nevertheless, although past experience indicates that public acceptance tends to increase after installation, visual intrusion is likely to remain an issue for some people.

Perhaps the concerns related to the public acceptance of wind farms, specifically, noise, the effect on birds, and visual intrusion in the landscape, are somewhat exaggerated. However, they will persist, especially as the concerns are partly subjective, and the wind energy industry will need to respond sensitively.

Technical development

There are many aspects of technical development that impinge on the prospects for wind power development in the UK. These include turbine evolution, turbine availability, wind variability, grid connection and radar, and electromagnetic interference. Each will be discussed briefly below. In addition, the challenge to adapt the technology to the offshore environment is considered.

The dominant type of wind turbine is the so-called Danish concept. It is a three bladed horizontal axis up-wind machine; that is, the rotor faces into the wind and rotates in a vertical plane. The generator is indirectly connected to the electrical grid through power electronics, thereby enabling the wind turbine rotor rotational speed to vary within some prescribed range. This facility is used to reduce the mechanical loads on the turbine and, in low wind speeds, to improve the aerodynamic efficiency and, most importantly, reduce aerodynamic noise. The blades can be pitched about their longitudinal axes to feather the blades and regulate the rotational speed of the rotor. For obvious reasons, this type of wind turbine is called a variable speed pitch regulated wind turbine.

Wind turbine technology has evolved rapidly over the last twenty years. The most obvious manifestation of this development is the exponential increase in machine size, see Fig. 5.5. In 1980, a large wind turbine had a rotor diameter of 20 m and a rated power, i.e. maximum generated power, of 50 kW. Today, a large wind turbine has a rotor diameter of 120 m and a maximum generated power of 5 MW.

In addition to becoming much larger, wind turbines have become more efficient. Since the wind speed varies, a wind turbine cannot produce rated power all

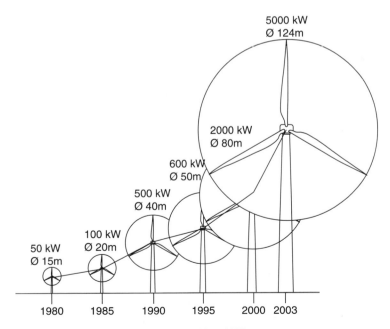

Figure 5.5 The growth in wind turbine size, 1980 to 2003.

the time. In the lowest wind speeds, the wind turbine is shut down since the power extracted from the wind is less than the internal losses. Above this lowest wind speed, the power produced increases with wind speed until the maximum power level of the wind turbine is reached. The power is then held constant at this maximum, except in the highest wind speeds when the wind turbine is again shut down. Consequently, the annual average power output from a wind turbine is much less than the rated power. A section of power generated by a 2 MW wind turbine in low wind speed is depicted in Fig. 5.6. The average power is measured by the capacity factor: the ratio of the average power to the rated power. In 1986, the typical target capacity factor for wind turbines was about 0.27 but the achieved capacity factor might be as low as half that value. The reason for this poor operational efficiency was the lack of reliability of some machines. Today wind turbines are much more reliable, attaining an availability in excess of 98%; that is, 98% of the time the turbines are capable of generating power provided the wind speed is not too low. On a good site, the capacity factor for a modern wind turbine is about 0.3 and sometimes even higher.

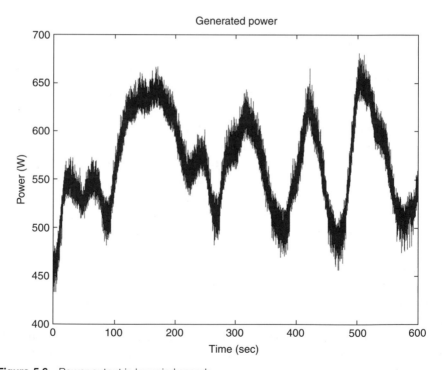

Figure 5.6   Power output in low wind speeds.

Because, as mentioned above, the wind speed varies causing the power output from a wind turbine to fluctuate, the value of wind power is sometimes questioned. It is argued that conventional generation with the same rating as the installed capacity of wind power, must be kept available in reserve to cover any deficit between electricity supply and demand. However, all electricity generation and supply systems require back-up including spinning reserve; that is, generators that are kept running and available to supply power at short notice. This reserve is necessary to cover any unexpected drop in electricity supply that might arise from a fault or loss of a large generation station. It is also required to cover any unexpected rise in electricity demand. The demand is partially predictable but an unpredictable residue that can be considered random, remains. On a local level, this residue may not be small but it is made much less significant by aggregation over the whole network. Similarly, the variability of wind power is much reduced by large-scale aggregation; that is, by having the wind farms distributed over a large geographical area. The aggregated wind power can again be considered as partially predictable, through exploiting weather forecasting information, with an unpredictable random residue. To accommodate wind power on an electricity supply network, it is treated as negative demand. The variability of demand is thus increased as it now includes the variability in the wind power. Sufficient generating back-up capable of covering this increased fluctuation in the electricity demand must be maintained. With the current very low level of wind power penetration in the electricity supply system, no increase in back-up is required. With a 10% penetration of wind power, only 300–500 MW of additional conventional back-up would be necessary. Even with all the costs of this extra back-up attributed to wind power, it is equivalent to adding only 0.3 c€/kWhour to the price.

The electricity supply system is not traditionally designed to accommodate generation distributed throughout the network, which would be the case with significant amounts of wind power. Instead, it is designed for large-scale central generation of power with outward transmission and distribution through the network. Nevertheless, the existing grid is expected to cope with 20% wind power penetration, although larger penetration would require its reconfiguration. In addition, there are concerns that wind power might reduce the stability of the grid, for example, through a fault propagating through the network causing wind turbines to serially disconnect. However, this issue is being addressed through the stipulation of appropriate grid connection codes for wind turbines. The wind turbine manufacturers are confident of their ability to meet these codes.

Wind turbines can have an adverse effect on communication systems and radar systems through electromagnetic interference. In communication systems, the transmitter or receiver must be in close proximity to the wind turbines. In radar systems, the interference is more serious but it can be countered by advanced filtering algorithms.

The UK has a rich offshore wind resource and it is expected that offshore development will be a major component of the UK's wind power development programme. The advantages of offshore development of wind power are considerable. The wind speeds are higher and the turbulence levels are lower than onshore. The visual intrusion, if not absent, is much less and there are no noise restrictions. In the absence of the latter, the wind turbines can be operated at higher rotor rotational speeds and so with lower loads. However, there are several disadvantages. There are higher capital costs because more substantial foundations are required offshore than onshore and because of connection by sub-sea electrical cable to the shore. Access to offshore wind turbines is restricted by poor weather conditions, in particular, strong winds or high seas. Consequently, operation and maintenance (O&M) costs are increased. As a fraction of the income of a wind farm, the O&M costs are approximately 10% to 15% onshore but are estimated to be 20% to 25% offshore. The technical challenge is to make offshore wind power more cost-effective by reducing the cost of O&M through improved reliability and proactive maintenance and by increasing yet further the size of wind turbines. The extent to which the latter can be achieved is not clear. Indeed, it may not be practical to make turbines much bigger than the existing maximum of 5 MW.

Concluding remarks

The price of wind power is almost competitive with conventional means of electricity generation. The UK has a rich exploitable wind power resource, particularly, towards the north and west of the country, but also round the south coast of England close to the main population centres. Although it is well established in Europe—and is making a significant contribution to electricity supply in Denmark, Spain, and Germany—wind power is still embryonic in the UK. Nevertheless, it is feasible that it could meet 10% of the UK's electricity demand. Such a large expansion would raise public concerns that include noise, impact on birds, and visual intrusion in the landscape. These concerns have a subjective element and are perhaps overstated, but large-scale development would need to be treated sensitively. To avoid difficulties over the public acceptance of wind power, offshore development would be preferable to onshore development and it is expected that offshore wind power will make a major contribution in the UK. In these circumstances, wind energy technology would enter a new era with many technological challenges. To conclude, wind power could make a major contribution to the UK's energy needs. The only major obstacle might be lack of political will.

Acknowledgement

Fig. 5.1 is reproduced with permission from the BWEA and Figs. 5.2 and 5.3 from Risø National Laboratory, Roskilde, Denmark.

Further Reading

General information on wind energy can be found in the following publications:

EWEA, 2004. *Wind Energy–The Facts–an Analysis of Wind Energy in the EU-25*, (available at http://www.ewea.org/index.php?id=91).

BWEA, 2005. *Top Myths about Wind Energy*, (available at http://www.bwea.com/energy/myths.html).

UK DTI, 2001, *The UK Wind Resource: Wind Energy Fact Sheet 8*, (available at http://www.dti.gov.uk/renewables/publications/pdfs/windfs8.pdf).

UK DTI, 2003, Our energy future—creating a low carbon economy, *Energy White Paper* (available at http://www.dti.gov.uk/energy/policy-strategy/energy-white-paper-2003/page21223.html)

An introductory text to wind energy technology is T. Burton, D. Sharpe, N. Jenkins, E. Bossanyi. 2004. *Wind Energy Handbook* , John Wiley & Sons Ltd., London.

Troen, I. and Peterson, E. L. 1989, *European Wind Atlas*, Riso National Laboratory, Roskilde.

The author

Bill Leithead is Professor of Control and Systems Engineering at the University of Strathclyde. He has been actively involved in wind energy both through academic research and collaboration with industry since 1988. He has published extensively on wind energy and has worked with many European manufacturers. Professor Leithead is currently on secondment to the Hamilton Institute in Ireland.

6. *Nuclear fission*

Sue Ion*

Introduction

This chapter will cover the nuclear fission option as a future energy supply, and will essentially address the question: can nuclear fission plug the gap until the potential of nuclear fusion is actually realized? (The potential for fusion is considered in detail chapter 7.) To put this question into context, let us first look at some of the key issues associated with nuclear fission, which currently supplies around one fifth of the UK's electricity.

The physics of fission

Most large scale power stations produce electricity by generating steam, which is used to power a turbine. In a nuclear power station, the principle is the same, but

---

* The author gratefully acknowledges the contribution and assistance of Adrian Bull in the preparation of this chapter.

instead of burning coal, oil, or gas to turn water into steam, the heat energy comes from a nuclear reactor. A reactor contains nuclear fuel, which remains in place for several months at a time, but over that time it generates a huge amount of energy. The fuel is usually made of uranium, often in the form of small pellets of uranium dioxide, a ceramic, stacked inside hollow metal tubes or fuel rods, which can be anything from a metre to four metres in length, depending on the reactor design. Each rod is about the diameter of a pencil, and the rods are assembled into carefully designed bundles, which in turn are fixed in place securely within the reactor.

There are two isotopes (or different types) of uranium, and only one of these is a material which is 'fissionable'—that is to say, if an atom of this uranium isotope is hit by a neutron, then it can split into two smaller atoms, giving off energy in the process and also emitting more neutrons. This, and other pathways, are illustrated in Fig. 6.1 (Source: CEA).

Controlling the reaction, so that the energy from the fission of uranium atoms is given out slowly over a period of years, requires two aspects of the process to be carefully balanced.

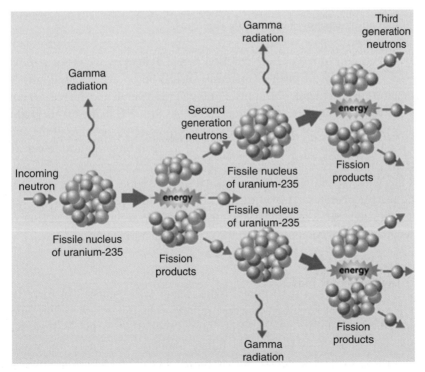

Figure 6.1 Schematic diagram of fission of ^{235}U

1. First, there must be enough fissile atoms in the fuel so that—on average— each fission leads to exactly one other. Any fewer, and the reaction will die away. Any more, and the reaction will speed up very rapidly and would—if not brought back under control—generate too much heat and damage the integrity of the fuel. The number of fissile atoms is controlled by 'enriching' the uranium to the right level when the fuel is fabricated. The fissile isotope of uranium (^{235}U) is just 0.7% of naturally occurring uranium, and this proportion is increased to around 3 or 4% during the enrichment process. Once enriched, the uranium is made into fuel pellets, as described above.

2. Second, the neutrons must be slowed down or 'moderated' if they are to cause fission to take place. If they are moving too fast, they essentially by-pass any uranium atoms they might come across. The moderation occurs as the neutrons pass through materials such as graphite or water. Different designs of reactor use one or other of these materials in the heart of the reactor core to slow the neutrons down. In a modern pressurized water reactor (PWR), the water also acts as the coolant—passing over the fuel to remove the heat. In older designs, such as Magnox or Advanced Gas-Cooled Reactors (AGRs), much of the reactor core is made up of graphite blocks, and the heat is removed by passing gas over the fuel rods.

Overall control of the nuclear reaction is maintained by means of removable 'control rods'. These are made of materials such as boron, which essentially 'soak up' neutrons in the reactor. The control rods can be lowered into the reactor core to slow down the reaction, or stop it all together. As the ^{235}U in the fuel is gradually used up over a period of months and years, the precise positioning of the control rods can be adjusted to compensate for this, by absorbing slightly fewer of the neutrons.

In this way, the uranium fuel provides a steady supply of heat to the coolant, as the reactor operates. This in turn is converted into electrical energy via the turbine. A single modern nuclear reactor can provide around 1,000 MW or more of electrical energy continuously to the power transmission grid.

The policy context

Back at the time when the Energy White Paper was launched in early 2003, the UK government identified four key pillars of energy policy. Environmental acceptability, reliability, affordability, and competition in the market. Never stated—but implicit—was a fifth pillar, namely the safety of whatever technologies might be deployed, both to the workforce and the wider public. The White Paper provided no overt support for nuclear fission, but did recognize the potential contribution it could make through its low carbon emissions. However, it also noted some issues associated with nuclear energy, in terms of the economics and in terms of waste, both of which, it was felt, needed resolution.

Cutting carbon emissions

In terms of the environment, the consideration which is perhaps the most important is the emission of CO_2, and Fig. 6.2 shows the comparison of nuclear with gas and coal in this respect.

The coal figure reflects coal without sequestration, and the figure for gas is for North Sea gas. If the UK's gas were to come from overseas—in whatever way—there would be additional energy associated with its transport, and thus additional emissions. Although these might not occur within the UK, they obviously still contribute to global climate change. Recognizing the UK's strong historical usage of coal, and the benefits it offers in terms of flexibility and reliability, we must consider generation from coal with capture and storage of the associated carbon dioxide as a key component of the future electricity mix. It is also important to note that in terms of CO_2 emissions, nuclear and wind power are actually rather similar. Both are dramatically lower than either coal or gas. At a time when CO_2 emissions are rising this is a vitally important consideration.

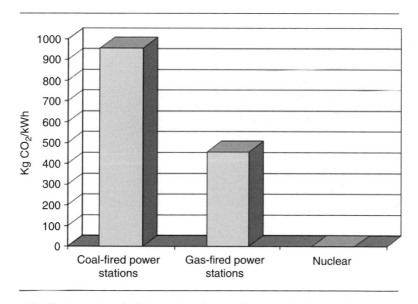

Figure 6.2 Typical carbon dioxide emissions from coal, gas, and nuclear stations.

Economics

When we look at how the economics of nuclear energy compare with other alternatives, there are inevitably a whole host of discussions to be had about key assumptions such as the rate of return which an investment must deliver, the length of time the plant will last, its performance, fuel costs, and so on. These apply not just to nuclear stations, but to the other options too, and it is unlikely that there will ever be definitive agreement on these issues between proponents of the various options, or indeed between energy market analysts. However, it is still instructive to look at the broad conclusions of recent authoritative studies, as these give a helpful indication of the relative economics.

One such study—looking at a whole range of figures from around the world, was carried out earlier this year by the Organization for Economic Development (OECD) in conjunction with the International Energy Agency (IEA) and the Nuclear Energy Agency (NEA)—see Fig. 6.3.

Importantly, this looked at a range of input assumptions and specifically looked at two different rates of return—5% and 10%. The chart clearly shows that nuclear is the cheapest option under either assumption, and whilst I am not going to argue that this in itself demonstrates nuclear would be the cheapest option in the UK, it is difficult to conclude that it would not be there or thereabouts in comparison with the leading fossil fuel alternatives. When we come to look in more detail at the ways in which we can cut carbon emissions, it is important to note also that nuclear is coming in at a much lower cost than renewables such as wind.

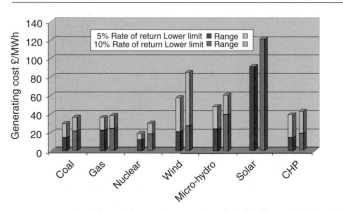

Source: "Projected Costs of generating Electricity", OECD; March 2005

Figure 6.3 Cost projections from OECD study: 'Projected costs of generating electricity', March, 2005.

Like the OECD study, the Royal Academy of Engineering concluded in 2004 that nuclear was very competitive in cost terms to gas, and much cheaper than renewables. This conclusion (illustrated in Fig. 6.4) was reached even before any allowance was made for the costs associated with carbon emissions. When those were included, the economics of nuclear were shown to be far and away the cheapest.

The pedigree of those studies illustrates that nuclear should not be discounted on the grounds of its economics. A further feature of nuclear energy relates to the stability of its economics. The cost of the finished fuel assemblies accounts for around 20% of the overall nuclear generating cost, and of this around a quarter—5%—is the cost of the raw uranium. The other 15% is associated with enrichment and fabrication costs. This is in stark contrast to, say, gas-fired generation, where the cost of the raw gas is typically 60% or so of the overall costs (see Fig. 6.5). That means that if the world market price of gas were to double (and that's by no means inconceivable, based on recent experience) the cost of gas-generated power would go up by 60% immediately. Yet if the world market price of uranium were to double, the impact on power costs from nuclear stations would be a mere 5%. Even this impact would not be felt for a year or so, when any newly purchased uranium had found its way to the power station in the form of fuel.

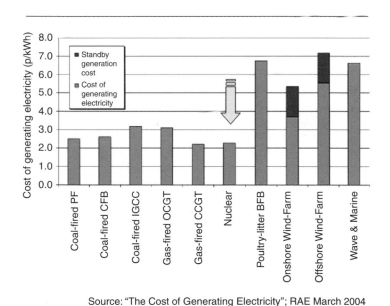

Source: "The Cost of Generating Electricity"; RAE March 2004

Figure 6.4 Cost projections from Royal Academy of Engineering Study: 'The cost of generating electricity', March, 2004.

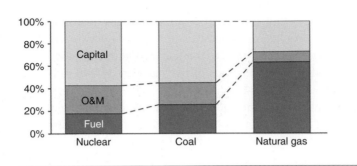

Figure 6.5 Relative contributions of capital, operation, and maintenance and fuel costs for different generation options.

Reliability of electricity supplies

But it is not just in terms of economics that nuclear offers reliability benefits. There is growing concern over the ability of two technologies to provide reliable power over the coming decades. These technologies are gas and renewables—in particular wind power—two forms which are projected to account for much of the UK's generating portfolio over future years.

Consider the following combination of issues which the UK is currently facing:

- The UK is in the midst of the 'dash for gas'—a huge shift in our power generation portfolio, which has seen us move from 65% coal-fired power in 1990, with no gas stations at all, to a position where we now expect at least two thirds of our power in 2020 to come from gas (see Fig. 6.6).

- The UK's nuclear fleet has already begun its progressive programme of station closures and is the first country in the world to be in this position, not due to any deficiencies in the fleet, but simply because the UK led the world in adopting nuclear in the first place. Fig. 6.7 shows how the next 10–15 years are likely to see most of the current fleet reach the end of its life.

- At the same time as the UK nuclear fleet is closing, it faces the prospect of losing a large tranche of our coal fired stations as well—due to the EU Large Combustion Plant Directive, which forces older coal plants either to close by 2015 or else to retro-fit equipment to clean up the discharges from such stations.

- Not only are global oil and gas reserves looking increasingly limited as demand increases, but the UK's domestic oil and gas reserves are well on

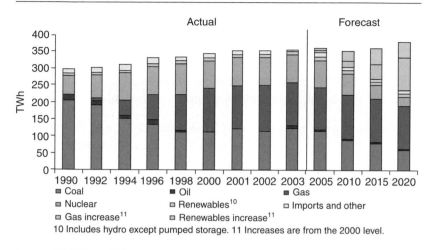

Source: DTI/Ofgem JESS report: November 2004

Figure 6.6 Historic and projected future power generation mix in the UK. (Source: DTI/Ofgem figures)

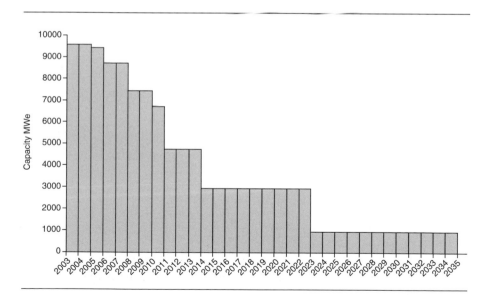

Figure 6.7 Projected fall in UK nuclear generation capacity.

the way to being depleted. The UK became a net importer of gas again in 2004, and as its reliance on gas rises, it could be importing up to 80% of its gas needs by 2020, by which time gas will be the dominant component of the mix. Such long term dependence on imported gas, with risks attached to both source and pipeline route, is a precarious situation for any nation. Disruption along any part of the route could put supplies in jeopardy.

There is an increasing emphasis being placed on both renewables and energy efficiency—both themes of the 2003 Energy White Paper. Both have a substantial role to play, but both present their own challenges:

- New renewable energy is mainly wind at present, and for the foreseeable future. This is inherently intermittent, and experience in countries such as Denmark and Germany, which have deployed substantial quantities of wind generation, has shown it to require substantial backup capacity if a reliable power supply is to be expected. The UK government has set an aspiration of 20% of the UK's electricity to come from renewables by 2020— a hugely ambitious objective. Yet even if this were to succeed, it would just offset the closure of carbon-free nuclear plants on approximately the same timescale, meaning that there would be no improvement in emission terms.

- Energy efficiency has—like renewables—a positive contribution to make, but huge step changes in demand as a result of energy efficiency measures should not be anticipated. A key limiting factor is that energy saving measures are most viable when fitted in new properties. Retro-fitting is usually much less cost effective. Yet the turnover rate of homes and commercial property is measured in decades, so rapid improvements are unlikely.

- The UK's electricity market is newly–liberalized and there is a need to gain confidence that the market will provide the necessary signals to prompt new investment in good time to meet demand.

- And finally unlike virtually all of its major economic competitors, the UK is essentially an island nation. There are very limited links to neighbouring nations for providing either raw fuels (coal, oil, and gas) or electricity via a network of power lines. The power link to France provides about 2 gigawatts worth of (largely nuclear) electricity.

So looking to the short and medium term future the UK is set for a period of substantial change and needs to consider how nuclear energy can contribute to the reliability of supplies.

First, and most obviously, looking at a future dominated by gas and renewables, anything else—including nuclear—contributes to increased diversity, which must of itself be helpful in reducing risk. Without new stations, nuclear will be providing just 3% of UK electricity in under 20 years time, compared with around 20% today.

Uranium is plentiful as a raw material, and comes from stable countries like Australia and Canada. Reliable baseload generation is a key feature of nuclear energy. Nuclear stations operate round the clock, day in and day out, irrespective of weather conditions. Apart from a periodic and predictable maintenance shutdown, the stations operate continuously at full power, providing the kind of baseload power a leading twenty-first-century economy demands.

Furthermore, it is highly credible to retain strategic stocks of nuclear fuel, just in case there ever were to be any sustained disruption to supply. The fabricated fuel to supply a fleet of ten new reactors (enough to supply 20% of UK electricity needs) for a year, would occupy only around 100 cubic metres. That means it would be small enough to fit comfortably within a very modestly-sized house. Finally—as mentioned earlier—nuclear provides valuable cost stability as well as supply reliability.

Potential new reactor technology

Let us now consider what technology would be available to the UK to move forward with nuclear. There are many options around the world at the moment. Interestingly, none of them have their genesis in the UK, unlike the systems currently operated which, with the exception of Sizewell B, are 'home grown' designs (see Table 6.1).

The most likely options for the UK, were new nuclear to go ahead, are the European Pressurized Water Reactor, which is a Franco-German product, and the Westinghouse AP1000, which is an American product. The EPR design has already been selected in two countries—in Finland, where it is already under construction, and in France. The AP1000 is the lead candidate in the US for delivery. Both products are in head-to-head competition in China for four units. China already has a significant amount of nuclear capacity, mainly based on French and Canadian technology. Several other options on the table are being built in the Asia-Pacific basin.

The EPR design is an improvement on the existing systems operating in Europe, with many safety features added, and enhanced protection against aircraft and earthquakes. These improvements were driven by the requirements of the European Utilities Requirements in the mid-1990s. It has some impressive technological advantages, but nevertheless is still considered proven, because it is

Table 6.1 Nuclear reactor technology currently under construction or consideration worldwide

Reactor Design	Type	Country of Origin	Lead Developer	Development Status
ABWR	BWR	US Japan	GE, Toshiba Hitachi	Operating in Japan. Under construction in Japan & Taiwan
CANDU-6	PHWR	Canada	AECL	Operating in Korea, China,
VVER-91/99	PWR	Russia	Atomstroyexport	Under construction in China
AHWR	PHWR	India	Nuclear Power Corporation of India	Starting construction
APR-1400	PWR	Korea, US	Kepco	Planned for Shin-Kori
APWR	PWR	Japan	Westinghouse & Mitsubishi	Planned for Tsuruga
EPR	PWR	France, Germany	Framatome ANP	Under construction in Finland Planned in France
AP1000	PWR	US	Westinghouse	Licensed in USA
SWR	BWR	France, Germany	Framatome-ANP	Offered in Finland
ESBWR	BWR	US	GE	Under development
ACR	PHWR	Canada	AECL	Under development

Figure 6.8 Impression of the EPR Plant currently under construction at Olkiluoto, Finland.

actually based on existing technology (the N4 design) that is operating in France today, and the Konvoi reactors in Germany.

Fig. 6.8 shows an impression of what the EPR will look like once completed in Finland, alongside the existing plants on the same site. The new reactor is due to be delivering power within five years. The French demonstrator of the same technology will follow soon after, with the expectation of a further fleet to follow in France.

The AP1000, the American competing product, is again designed to meet the energy markets of the twenty-first century. It embodies safe, passive systems—with a lot of simplicity in the design, which is key to the economic benefit. In terms of economics, it is certainly competitive with the latest CCGT gas technology. Again—like the EPR—AP1000 uses proven components such as the plant's steam generators, the pressure vessel and the pumps are already in operation in different reactors worldwide (see Fig. 6.9).

On the economics, one might well ask how confident can the industry be that they will be competitive. We know how many components and how much material in terms of concrete, pumps, valves, and so forth, went into Sizewell B. And in comparison we know how many of the same items will go into a new design such as an EPR or AP1000. The difference, shown in Fig. 6.10 for the AP1000, is striking—largely due to the passive safety systems and associated simpler design. It is easy to see that the parts list is much shorter, so there's a very high degree of confidence in the plant being markedly cheaper than a conventional design.

- Designed for the energy markets of the 21st Century.
- Safe
  - Passive systems improve safety
- Simple
- Passive safety simplifies design, operation and maintenace
- Economic
- AP1000 competitive with CCGT
- Use proven components already in operation

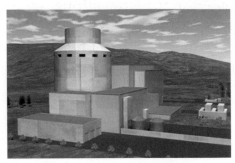

Figure 6.9   Impression of the Westinghouse AP1000 reactor.

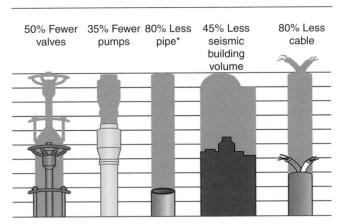

Compared with a conventional 1000 MW PWR

Figure 6.10   Components in the AP1000 compared with a conventional design.

As well as benefits arising from the actual design philosophy of the reactor, there are improvements to be anticipated from the way construction is approached in modern nuclear reactor systems. Whichever design might be selected, an increasing proportion of the construction nowadays takes place remote from the reactor site itself, with sub-modules of the building being fabricated in factories and then shipped whole to the site for assembly. In this way, a large number of construction activities can be carried out in parallel and the on-site construction work is kept to a minimum. Quality control is also often easier in a factory setting than on the construction site.

Investor considerations—regulation

So with all the benefits nuclear can offer to policy makers, and with a range of suitable technologies available, why is nuclear energy shunned in the UK as we speak? A variation on this question is often used by ministers and their teams to head off the nuclear debate. The line is basically: if nobody from the private sector is beating a path to my door wanting to build a nuclear reactor, why do we need to think about whether or not it's a 'good thing'?

The fact is that something of a 'Catch 22' situation has been created. Although a new nuclear plant would only take four or five years to build, there are the planning and approvals processes to go through in the same way as for any other form of energy. However, there is also a three-year hurdle to overcome that is unique to nuclear energy, namely the need to get the reactor design approved by the Nuclear Installations Inspectorate, the UK's nuclear regulator.

There has previously been a strong tendency for UK regulators to request changes to make a design 'better' or 'more familiar' with the result that it becomes a unique station, with a lot of UK add-ons. This approach has led to the UK fleet being enormously varied (for instance virtually all the Magnox stations were different, and each reactor design used a slightly different design of fuel from the other stations). Such an approach has led to a missed opportunity in terms of economies of scale during operation in contrast to, for instance, the French who have a series of near-identical fleets (they have over thirty 900 MW plants all similar to one another, and more than twenty 1300 MW units, all of very similar design). More importantly in the context of new build investment, the UK approach implies a design which is still being changed as the licensing progresses, which has clear potential to introduce delays.

The way forward is to adopt proven international reactor systems and to build them according to standard designs. On that basis, and with a streamlined—but not shortcut—planning and regulatory regime (where, for instance, the Public Inquiry focuses just on local issues) a potential predictable timeline for delivery would be as shown in Fig. 6.11.

Investor considerations—delivery and operational performance

Korea and Japan provide many examples of construction and commissioning performance on time and within budget.

Fig. 6.12 shows Tepco's biggest site, at Kashiwazaki-Kariwa on the sea of Japan. It houses seven large boiling water reactors, delivering 8.2 gigawatts of

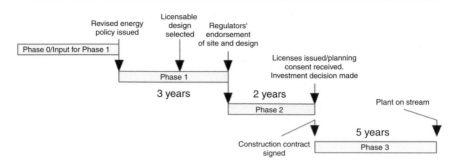

Figure 6.11   Potential timeline for a new nuclear plant in the UK.

Figure 6.12   The Japanese nuclear reactor site at Kashiwazaki-Kariwa.

electricity, 80% of Tokyo's power demand, day in, day out. It is a very successful example of the benefits of series build.

Another success story is the upwards trend in capacity factors of nuclear plants, as a result of sensible approaches to operating and maintenance and to standardization of fleets. Levels have risen steadily and the best stations in the world routinely achieve well over 90% (see Fig. 6.13). Any utility investing in a new reactor now would be really disappointed if they didn't get at least a 90% capacity factor out of their new unit.

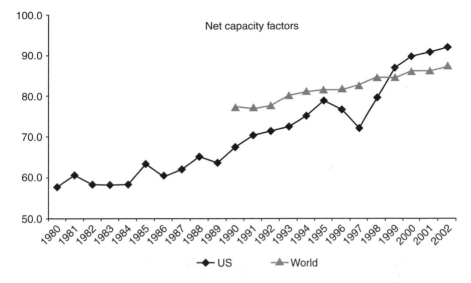

Source: WAN O and Nuclear Energy Institute

Figure 6.13   Improvements in nuclear capacity factor.

Investor considerations—waste

It is important now to look at a very emotive issue for many people namely the thorny issue of nuclear waste. It is most important to see this issue in context. There is a significant nuclear waste inventory in the UK. Most of that legacy is associated with the way in which we went about development of nuclear energy back in the 1940s, 50s and 60s, and any decision we make about new nuclear stations will not change that fact. Nor will it change the fact that those wastes are being managed safely and effectively.

Waste associated with new reactors would only add a modest amount to this inventory. If the UK were to build a new series of reactors that would replace its capacity and run for 60 years, Fig. 6.14 shows the additional waste that would be generated compared with that which already exists at sites such as Sellafield, Dounreay, and the operating power stations. So a new build decision—even on the scale of a whole replacement fleet—would add only around one tenth to the volumes of existing wastes. Furthermore, the new build wastes would not bring any new technical challenges to add to those already faced. In fact the reverse is true, modern stations are designed with waste management and decommissioning in mind, and so the wastes are much easier to deal with than some of the challenges presented by the older legacy materials.

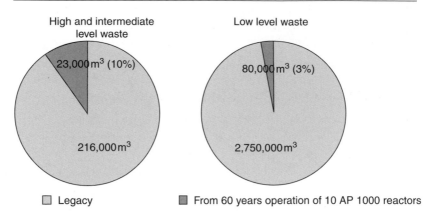

Figure 6.14   Waste arising from potential new build nuclear planets in comparison with the existing UK nuclear waste legacy. Wastes from new stations would make only a small addition to existing stocks.

As for the reasons behind the dramatic differences in waste volumes between future and past systems, much has to do with the history of the older Magnox stations. These reactors were the workhorses—the very good workhorses—of the UK fleet for many years, and simply had a lot more waste associated with their operation than modern systems. They were relatively large and complex systems considering the amount of power produced, and fuel requirements were notably higher in terms of tonnes of fuel needed each year, as the burn-up levels achieved were far lower than later designs. Another factor is the fact that the industry carried out a lot of development work on sites like Sellafield in the 1940s, 50s, and 60s in order, for instance, to build the unique radiochemical processing facilities which are based there.

That deals with the amounts of waste, but the concern to investors is not simply the quantities of material, but the process for dealing with it. The UK currently has no formal policy on management of the higher level wastes, so investors in a potential new plant simply don't know what they will be required to do with their wastes at some point in the future. Until a clear policy is put in place (and only governments can do this) then this policy vacuum is essentially an open-ended liability for the investor.

In terms of waste policy, what investors need to see is certainty. In the UK the Committee on Radioactive Waste Management (CoRWM) made recommendations in July 2006, which were accepted by the Government and welcomed by the industry. Progress should now begin towards selection of disposal sites. We already have a disposal route for low activity waste. Other countries have successfully addressed the issue of higher level wastes—Finland,

Sweden, and the USA have all identified the way to proceed, with Switzerland and Japan moving that way. The issues are not technical, but political, in terms of agreeing the process and deciding where the final disposal site is to be.

Investor considerations—the electricity market

In terms of an investor view of the electricity market and revenue issues, the long timescales are certainly an issue. As noted earlier, it would be five years from the announcement of a more positive policy until the likely date of a final investment decision. A further five years from that investment decision to the first date of power generation, and then an operating lifetime of perhaps 60 years. That represents a remarkably long time over which to project likely revenues and over which to judge the likelihood of any policy changes. Given the recent level of major change and intervention in the electricity market, it is perhaps not surprising that potential investors in nuclear are not rushing to place their money in nuclear projects. An investor looking at an outlay of perhaps £2 billion would have to take a very deep breath before making that sort of commitment in the absence of any reassurance from government that the goal posts are not going to move in some unpredictable and unhelpful way.

Fuel and power prices are showing increasing volatility of late, and that works in favour of the nuclear industry, because of the price stability discussed earlier. But the energy sector in general—and nuclear issues in particular—are highly political. Governments can take a politically-motivated decision to phase out or to stop at any moment in time, and that knowledge also affects investor confidence. Public recognition by the government of the role that nuclear could play in delivering energy policy benefits (in particular with regard to climate change) would go some way towards reassuring investors that policy will not be changed 'on a whim'.

Public attitudes

Moving now from investor perspectives to look at public attitudes, we find that, with respect to nuclear, these can be very ambivalent. Fig. 6.15 illustrates one important question that was asked early in 2005.

The response was that 35% of people were supportive, 30% opposed, but many people did not have a strong view. That is a major consideration because the 'don't knows' are very easily swayed by an incident of the day, or the feeling of the day. However, the fact that more people were prepared to express a supportive opinion than a negative one on the issue is and contradicts what certain news media will suggest is the general flavour of public attitudes towards nuclear energy.

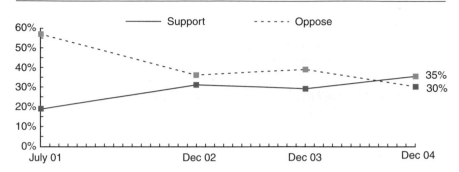

Figure 6.15   Recent UK public opinion survey concerning attitudes towards nuclear.

Those figures reflect the views of the general public as a whole. But if there were to be new stations built in the UK, they would almost certainly be built on existing sites where nuclear power stations are already operating, or are being decommissioned. When we survey the opinion around those existing sites, then we find a much more favourable response to whether or not these stations should be built. That again is encouraging, as people in these communities recognize that there is a station near them, and very often take the trouble to look below the surface and find out the reality for themselves about the issues surrounding nuclear generation in much more detail.

The longer term future

The reactor technology options currently available have been outlined above, along with the argument in favour of building these to replace some of our existing power generation fleet. But looking at the bigger picture and thinking in the longer term, it becomes clear that more than just that is needed to make major cuts in the level of our carbon emissions. The Energy White Paper committed the UK government to achieving a 60% cut in emissions (relative to 1990 levels) by the year 2050. This was a level recommended by the Royal Commission on Environmental Pollution, in order to keep the threat of climate change in check. Yet a look at the UK's overall energy usage as a nation in Fig. 6.16 soon illustrates the scale of that task.

This chart of UK energy consumption in million tonnes of oil equivalent, shows that only a fraction of total energy use is consumed in the form of electricity. Even if we had no carbon emissions at all from the electricity sector, it would not even be close to a 60% cut in overall emissions when the coal, oil, and gas used elsewhere were considered.

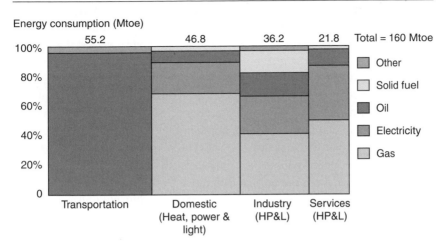

Figure 6.16   Breakdown of total UK energy demand.

One major initiative looking long term is an international programme called Generation IV. It is designed to reflect a 'middle of the twenty-first century' vision of energy needs, and to provide a pathway to systems that could start to operate between 2020 and 2040 and beyond. These are designed with sustainability, safety, reliability, and economic goals in mind, so they will offer a number of attractive features. They will maximize the energy we get out of the resources we put in. They will obviously be carbon-free, helping to combat climate change, whilst also minimizing future wastes. They will help to prevent proliferation, and ensure maximum physical protection, safety, and reliability. They are designed to give very low likelihood of unplanned events, and to eliminate the need for off-site emergency responses, so offer very high degrees of safety. In terms of economics, they give lifecycle cost advantages over other sources, and are generally small and modular rather than 'big and beautiful', because smaller systems often have advantages in deregulated markets.

One key concept to highlight as an example, is the high temperature gas cooled reactor (HTR), and more specifically the pebble-bed design. There is an HTR operating now in Beijing, delivering 10 megawatts to their grid, and in South Africa there is an international consortium that will build a commercial pebble-bed unit of 180 megawatts at Koeberg. The aim is to have that plant coming online around 2011, as a precursor to the longer-term journey. As mentioned above, these plants are modular, inherently safe, and they are highly suited to cost effective hydrogen production because of the high temperatures at which they operate. The outlet temperature in this design is about 950°C. Some future systems are looking at higher temperatures still.

Such temperatures offer the opportunity to do things differently, for instance opening up more direct—and therefore more efficient—routes to hydrogen production. Hydrogen is often talked about as an energy vector that can help the energy pattern in the world, but it has to be produced from carbon-neutral sources. Light water reactors, like any other form of power generation, can generate electricity which is used to produce hydrogen by electrolysis of water. Similarly, carbon-free power for such a process could also come from wind and other renewables. That is well proven, but is not the most economic or effective way to make hydrogen. But once we have nuclear plants which use heat in the high temperature range, then we have the opportunity to couple electricity generation with the heat output from the unit, and to move to thermally assisted 'cracking' of water to produce the hydrogen. Still carbon-free, but now also much more efficient, and that becomes increasingly important when we consider just how much hydrogen production will need to take place in future decades to underpin that shift in our transport systems.

This is just one example of how the nuclear industry of tomorrow can offer much more than merely power generation.

So, to summarize, the UK has opportunities select from internationally available, standardized systems, that are being built or considered elsewhere. In doing so, the UK can expect the same economic benefits that have driven countries such as Finland to make the same choice. The expectation is for reliable operation over 60 years at 90% capacity factor, and maybe even more.

Looking towards the time that fusion might be available, there will be further choice of these more advanced systems which will be being deployed worldwide. They are not UK-unique systems. But will come from international consortia and deliver power to the UK grid on a routine basis, in the same way as France already gets most of its electricity routinely from nuclear energy.

The author

Dr Sue Ion is an independent consultant in the Energy Sector having spent 26 years with BNFL and 10 years as its Chief Technology Officer. She is the UK representative on the IAEA Standing Advisory Group on Nuclear Energy and the Euratom Science and Technology Committee. She is also a member of the UK Council for Science and Technology and the EPSRC Council. Between 2004 and 2006 she was President of the British Nuclear Energy Society. She has spoken widely on matters associated with the nuclear industry, is a Visiting Professor at Imperial College, and a Member of the Board of Governors of the University of Manchester.

7. *Fusion energy*

Chris Llewellyn Smith and David Ward

Introduction

Fusion powers the Sun and stars, and is potentially an environmentally responsible and intrinsically safe source of essentially limitless energy on earth. Experiments at the Joint European Torus (JET) in the UK, which has produced 16 MW of fusion power, and at other facilities, have shown that fusion can be mastered on earth.

Fusion power is still being developed, and will not be available as soon as we would like. We are confident that it will be possible to build viable fusion power stations, and it looks as if the cost of fusion power will be reasonable. But time is needed to further develop the technology in order to ensure that it would be reliable and economical, and to test in power station conditions the materials that would be used in its construction.

Assuming no major surprises, an orderly fusion development programme— properly organized and funded—could lead to a prototype fusion power station putting electricity into the grid within 30 years, with commercial fusion power following some ten or more years later. A fusion power station is effectively a tiny 'artificial sun'.

Principles of fusion

Reactions between light atomic nuclei in which a heavier nucleus is formed with the release of energy are called fusion reactions. The reaction of primary interest as a source of power on Earth involves two isotopes of hydrogen (Deuterium and Tritium) fusing to form helium and a neutron:

$$D + T \rightarrow {}^4He + n + \text{energy (17.6 million electric volts [MeV])} \qquad (7.1)$$

Energy is liberated because Helium-4 is very tightly bound: it takes the form of kinetic energy, shared 14.1 MeV/3.5 MeV between the neutron and the Helium-4 nucleus (a chemical reaction typically releases $\sim$1 eV [electron volt], which is the energy imparted to an electron when accelerated through 1 volt).

To initiate the fusion reaction (1), a gas of deuterium and tritium must be heated to over 100 million°C (henceforth: M°C)—ten times hotter than the core of the Sun. At a few thousand degrees, inter-atomic collisions knock the electrons out of the atoms to form a mixture of separated nuclei and electrons known as a plasma. Being positively charged, the rapidly moving deuterons and tritons suffer a mutual electric repulsion when they approach one another. However, as the temperature—and hence their velocities—rises, they come closer together before being pushed apart. When the temperature exceeds 100 M°C, the more energetic deuterons and tritons approach within the range of each other's nuclear force and fusion can occur copiously.

There are two challenges. The first is to heat a large volume of deuterium plus tritium (D and T) gas to over 100 M°C, while preventing the very hot gas from being cooled (and polluted) by touching the walls: as described below, this has been achieved using a 'magnetic bottle' known as a tokamak. The helium nuclei that are produced by fusion (being electrically charged) remain in the 'bottle', where their energy serves to keep the gas hot. The neutrons, however, are electrically neutral and escape into, and heat up, the walls: this heat is then used to drive turbines and generate electricity.

The huge flux of very energetic neutrons and heat (in the form of electromagnetic radiation and plasma particles) can damage the container. This leads to the main outstanding challenge, which is to make a container with walls sufficiently robust to stand up, day-in day-out for several years, to this neutron bombardment and heat flux.

Fusion fuel

The tiny amount of fuel that is needed is one of the attractions of fusion. The release of energy from a fusion reaction is ten million times greater than from a typical chemical reaction, such as occurs in burning a fossil fuel. Correspondingly, whereas a 1 GW coal power station burns 10,000 tonnes per day of coal, a 1 GW fusion power station would burn only about 1 kg of D+T per day.

Deuterium is stable, and in one in every 3,350 molecules of ordinary water one of the hydrogen atoms is replaced by a deuterium atom. Deuterium can be easily, and cheaply, extracted from water. Tritium, which is unstable and decays with a half-life $\sim$12 years, occurs only in tiny quantities naturally. But, as described below, it can be generated in-situ in a fusion reactor by using neutrons from the fusion reaction impacting on lithium to produce tritium in the reaction:

$$\text{Neutron} + \text{Lithium} \rightarrow \text{Helium} + \text{Tritium.} \qquad (7.2)$$

The raw fuels of a fusion reactor would therefore be lithium and water. Lithium is a common metal, which is in daily use in mobile phone and laptop batteries. Used to fuel a fusion power station, the lithium in one laptop battery, complemented by deuterium extracted from 45 litres of water, would (allowing for inefficiencies) produce 200,000 kW-hours of electricity—the same as 70 tonnes of coal: this is equal to the UK's current per capita electricity production for 30 years.

Fusion power stations

Fig. 7.1 shows the conceptual layout (not to scale) of a fusion power station. At the centre is a D–T plasma with a volume $\sim$1000 m^3 (actually contained in a 'toroidal' [doughnut shaped] chamber—see below). D and T are fed into the core and heated to over 100 M°C, a temperature routinely achieved at JET, as described below. The neutrons produced by the fusion reaction (1) escape the magnetic bottle and penetrate the surrounding structure, known as the blanket, which will be about 1 metre thick.

In the blanket, the neutrons encounter lithium and produce tritium through reaction (2). There are various competing reaction channels, which do not produce tritium directly, but some of them produce additional neutrons that can then produce tritium (the production of additional neutrons can be enhanced, for example, by adding beryllium or lead). The upshot is that, on paper at least, it is possible to design fusion reactors that would produce enough tritium for their own use plus a small surplus to start up new plants: this will be tested at ITER (the International Tokamak Experimental Reactor), as described below.

The neutrons will also heat up the blanket, to around 400°C in so-called 'near-term' power plant models that would use relatively ordinary materials, and up to

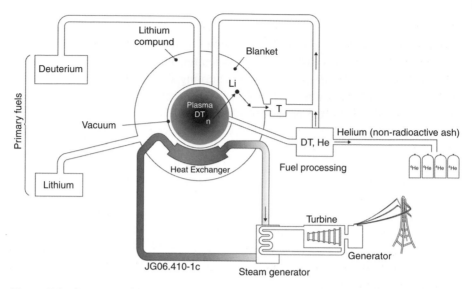

Figure 7.1 A conceptual fusion power station is similar to an existing thermal power station but with a different furnace and fuel. The figure is not to scale; in reality the fusion core would be a very much smaller part of the whole power station, and the 'blanket' would be ~1 m thick while the plasma (which, as explained later in the text, would be contained in a toroidal chamber) would occupy ~1000 m^3.

perhaps 1100°C in models that use advanced materials such as silicon carbide. The heat will be extracted through a primary cooling circuit, which could contain water or helium, that in turn will heat water in a secondary circuit that will drive turbines.

Attributes of fusion

The advantages of fusion are:

- no CO_2 or air pollution;
- essentially unlimited fuel;
- intrinsic safety;
- 'internal' costs (i.e. costs of generation) look reasonable—see the discussion of power plant studies below ('external' cost—impact on health, climate, and the environment—will be essentially zero);
- it will meet a vital need.

There is enough deuterium for millions of years, and easily mined lithium for several thousand years (after which it could be extracted from water).

A key safety feature is that, although it will occupy a large volume, the amount of tritium and deuterium in a fusion reactor will be tiny: the weight of the hot fuel in the core will be about the same as ten postage stamps. Because the gas will be so dilute, there will be no possibility whatsoever of a dangerous runaway reaction. Furthermore, there is not enough energy inside the plant to drive a major accident and not much fuel available to be released to the environment if an accident did occur.

What are the hazards? First, although the products of fusion (helium and neutrons) are not radioactive, the blanket will become activated when struck by the neutrons. However, the radioactivity decays away with half-lives of order ten years, and all the components could be recycled within 100 years. Should the cooling circuit fail completely, radioactivity in the walls would continue to generate heat, but the temperature would peak well below the temperature at which the structure could melt.

Second, tritium is radioactive, but again the half-life is relatively short (12 years) and the possible hazard is not very great. In any case it will be easy to design reactors so that even in the worst imaginable accidents or incidents (such as earthquakes or aircraft crashes) only a small percentage of the tritium inventory could be released and evacuation of the neighbouring population would not be necessary.

The current status of fusion research

The most promising magnetic configuration for confining ('bottling') fusion plasmas is called a tokamak (a contraction of a Russian phrase meaning toroidal chamber with a magnetic coil). The basic layout of a tokamak is shown in Fig. 7.2.

An understanding of how tokamaks work is not needed to follow the rest of this paper, but for readers with a technical background who are interested:

- A small amount of gas (hydrogen or deuterium in most experiments; deuterium and tritium in some experiments at JET and in an actual fusion reactor) is injected into the toroidal (doughnut shaped) vacuum chamber after the magnetic field coils have been switched on.

- A current is driven in a coil wound around the column at the centre, which acts as the primary of a transformer. This drives an electric current ($\sim$5 MA in JET) through the gas, which acts as the secondary.

- The electric current heats the gas, and turns it into a plasma. It also produces a magnetic field which, combined with the magnetic field produced by the

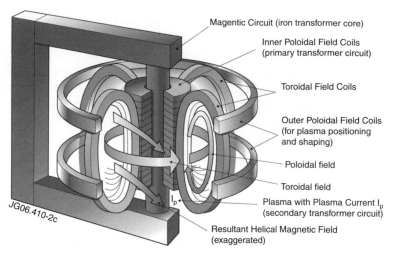

Magentic Circuit (iron transformer core)

Inner Poloidal Field Coils
(primary transformer circuit)

Toroidal Field Coils

Outer Poloidal Field Coils
(for plasma positioning
and shaping)

Poloidal field

Toroidal field

Plasma with Plasma Current I_p
(secondary transformer circuit)

Resultant Helical Magnetic Field
(exaggerated)

JG06.410-2c

I_p

Figure 7.2   In a tokamak, the fusion fuel is held in a toroidal chamber surrounded by magnets. A current is induced in the fuel by transformer action and, together with the magnets, produces a helical magnetic structure that holds the hot fuel away from the wall.

external coils, serves to 'confine' the plasma, i.e. hold it away from the walls and provide very good thermal insulation.

- The current induced by transformer action can only heat the plasma to about one third of the temperature needed for copious fusion to occur. Additional heating power must therefore be supplied, by mechanisms that serve also to drive the current (which is essential for plasma confinement) thereby keeping it flowing.

- This additional heating and 'current drive' can be provided by injecting either microwaves (rather as in a microwave oven) or beams of very fast, energetic neutral particles, produced by banks of small accelerators, which transfer energy to the plasma through collisions, or both. Many MWs of heating power can be supplied by these means.

In addition to heating and current drive systems, tokamaks are equipped with 'diagnostic' devices that measure the magnetic field, electron and ion temperatures and densities, the plasma pressure position and shape, neutron and photon production, impurities etc., and monitor the development of instabilities

Three parameters control the fusion reaction rate:

1. The plasma temperature (T), which as already stated must be above 100 M°C.

2. The plasma pressure (P). The reaction rate is approximately proportional to P^2.

3. The 'energy confinement time' (τ_E) defined by:

$$\tau_E = \frac{\text{energy in the plasma}}{\text{power supplied to heat the plasma}}$$

τ_E measures how well the magnetic field insulates the plasma. It is obvious that the larger τ_E, the more effective a fusion reactor will be as a net source of power.

It turns out that the 'fusion product' P (in atmospheres) $\times$ τ_E (in seconds) determines the energy gain of the fusion device, and this must be ten or more in a fusion power station. The 'fusion performance plot' (Fig. 7.3) of Pτ_E vs. T, which shows data points from different tokamaks, indicates the substantial progress towards power station conditions that has been achieved in recent decades.

Semi-empirical scaling laws have been devised that interpolate rather accurately between results from machines with very different sizes, magnetic fields and plasma currents. The scaling law for confinement time is shown in

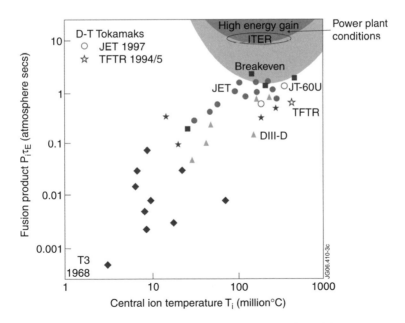

Figure 7.3 Selected results from different tokamaks demonstrate substantial progress over recent decades from the low temperature, low energy gain points at the bottom left. Temperatures above 100 M°C are now routinely achieved and an energy gain of around one has been reached. A power plant needs an energy gain above ten and this should be achieved in ITER.

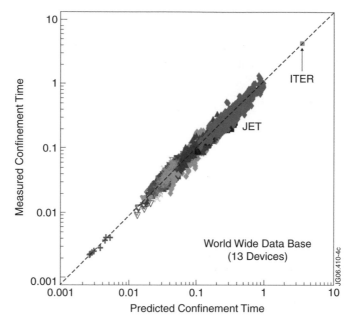

Figure 7.4 Confinement times (in seconds) measured at a range of very different tokamaks are well described by a semi-empirical scaling law that 'predicts' the confinement time as a function of the tokamak's size, magnetic field and plasma current (and other parameters). The prediction for ITER is shown.

Fig. 7.4. This figure makes us rather confident that the power station sized device called ITER, which (see below) will be constructed in the near future, will perform as advertised. ITER should confirm that it is possible to build a fusion power station (provided that, meanwhile, work on the materials that will be used to construct a power station proceeds apace, as described below).

JET is the closest existing tokamak to ITER in size and in performance (in many, if not all, parameters). It is also the only tokamak in the world that can be operated with tritium, although most of the time Deuterium alone (which has essentially the same plasma properties as a D—T mixture, but a much lower fusion cross-section) is used—as in other tokamaks—in order to keep activation to a minimum. JET holds the world record for fusion energy and power production in the two pulses shown in Fig. 7.5, which also shows the record pulse from the now closed Tokamak Fusion Test Reactor (TFTR) at Princeton, which is the only other tokamak to have been capable of using tritium. The record energy pulse produced was ~4 MW for five seconds, after which the pulse was ended in order to keep within the neutron production budget. Record power, of 16 MW, was produced when trying to push up the plasma pressure—but this could only be

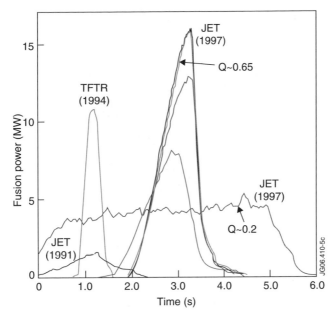

Figure 7.5 Fusion power produced in record pulses in D-T plasmas at JET and TFTR. Here Q is the fusion power generated/heating power, which is expected to be at least 10 at ITER.

sustained for a short time. The predicted performance of ITER is based on the lower power, steady plasmas, although as discussed below, one of the major goals of ongoing tokamak research is to find ways of pushing up the pressure (and hence the fusion reaction rate, which P^2) to higher values in a controlled way.

The next steps—ITER and IFMIF

Two intermediate facilities are necessary (which, see below, can and should be built in parallel) before the construction of a prototype fusion power station, fully equipped with turbines etc., will supply power to the grid. These are:

ITER (International Tokamak Experimental Reactor)

ITER, which is shown in Fig. 7.6, will be approximately twice the size of JET in linear dimensions, and operate with a higher magnetic field and current flowing through the plasma. The aim of ITER is to demonstrate integrated physics and engineering on the scale of a power station. The design goal is to produce 500 MW of fusion power, with an input ∼50 MW.

JET can only operate for up to one minute because the toroidal coils that produce the major component of the magnetic field are made of copper and heat up. This would not be acceptable in a fusion power station, and ITER will be equipped with

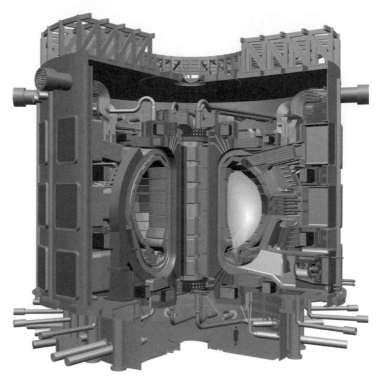

Figure 7.6 The ITER project, ready for construction, is designed to produce at least 500 MW of fusion power. It is similar in configuration to JET but twice as large (in each dimension).

super-conducting coils, that will allow indefinite operation if the plasma current can be kept flowing (the design goal is above ten minutes). Super-conducting tokamaks already exist, and others are under construction, but super-conducting coils have not so far been used in really large tokamaks capable of using tritium. ITER will also contain test blanket modules that, for the first time, will test features that will be necessary in power stations, such as, for example, the in situ generation and recovery of tritium.

A major goal of ITER is to show that existing plasma performance can be reproduced with much higher fusion power than can be produced in existing devices. Developments with the potential to improve the economic competitiveness of fusion power will also be sought (in experiments at existing machines as well as ITER). The main goals are:

1. Demonstrating that large amounts of fusion power (10 times the input power) can be produced in a controlled way, without provoking uncontrolled instabilities, over-heating the surrounding materials or compromising the

purity of the fusion fuel. These issues are successfully managed in existing devices but will become much harder at higher power levels produced for longer times. ITER is designed to tolerate this but it remains a big challenge.

2. Finding ways of pushing the plasma pressure to higher values (the rate at which fusion occurs and produces power, is proportional to the square of the pressure) without provoking uncontrollable instabilities. This would allow a power plant to operate either at higher power density or with reduced strength magnets, in either case lowering the expected cost of fusion generated electricity.

3. Demonstrating that continuous ('steady state') operation, which is economically and technically highly desirable if not essential, can be achieved without expending too much power. There is optimism that the plasma current can be kept flowing indefinitely by 'current drive', from radio-frequency waves and particle beams, boosted by a self generated ('bootstrap') current, however this must be optimized to minimize the cost in terms of the power needed.

Prototypes of all key ITER components have been fabricated by industry and tested. ITER, which will cost €4.5 billion, will be funded and built by a consortium of the European Union, Japan, Russia, USA, China, South Korea, and possibly India. Construction will begin, at Cadarache in France, once planning permission—which is being sought at the time of writing—is granted.

IFMIF (International Fusion Materials Irradiation Facility)

Those 'structural' materials, from which fusion power stations will be built, that are close to the plasma will be subjected to many years of continuous bombardment by a $\sim 2.5\,\text{MWm}^{-2}$ flux of 14 MeV neutrons. This neutron bombardment will on average displace each atom in nearby parts of the blanket and supporting structures from its equilibrium position some 30 times a year. Displaced atoms normally return to their original configuration, but occasionally they do not and this weakens the material. On the basis of experience of neutron-induced damage in fast breeder reactors, it seems that materials can be found that would meet the target of a useful lifetime of around five years before the materials would have to be replaced. The much higher energy fusion neutrons will, however, initiate nuclear reactions that can produce helium inside the structural materials, and there is a concern that the helium could accumulate and further weaken them. The so-called plasma-facing materials and a component called the divertor (through which impurities and the helium 'ash' produced in D-T fusion are exhausted) will be subjected to additional fluxes of $500\,\text{kWm}^{-2}$ and

$10 \, \mathrm{MWm^{-2}}$ respectively in the form of plasma particles and electromagnetic radiation. Special solutions are required and have been proposed for these areas, but they need further development and testing in reactor conditions.

Various materials are known that may be able to remain robust under such bombardments (it is in any case foreseen that the most strongly affected components will be replaced periodically). However, before a fusion reactor can be licensed and built, it will be necessary to test the materials for many years in power station conditions. The only way to produce neutrons at the same rate and with essentially the same distributions of energies and intensity as those that will be experienced in a fusion power station, is by constructing an accelerator-based test facility known as IFMIF (International Fusion Materials Irradiation Facility). Further modelling and proxy experiments (for example, using fission and neutrons produced by spallation sources) can help identification of suitable candidate materials. But they cannot substitute for IFMIF, and neither will testing in ITER be sufficient, because (i) the neutron flux will only be $\sim$30% that in an actual fusion power station, in which the fusion power will be several GWs, and (ii) as an experimental device, ITER will only operate for at most a few hours a day, while IFMIF will operate round the clock day-in day-out.

IFMIF, which will cost $\sim$ €800 M, will consist of two 5 MW accelerators that will accelerate deuterons to 40 MeV (very non-trivial devices). The two beams will hit a liquid lithium target that will produce neutrons, stripped out of the deuterons, with a spread of energies and an intensity close to that generated in a fusion reactor. These neutrons will provide estimated displacement rates (in steel) of 50, 20, and 1 displacements per atom per year over volumes of, respectively, 0.1, 0.5, and 6 litres.

Power plant studies

A power plant conceptual study has recently been completed under the auspices of the European Fusion Development Agreement. This study provided important results on the viability of fusion power, and inputs to the critical path analysis of fusion development described below.

Four models (A–D) were studied as examples of a spectrum of possibilities. Systems codes were used to vary the designs, subject to assigned plasma physics and technology rules and limitations, in order to produce an economic optimum. The resulting parameterization of the cost of fusion generated electricity as a function of the design parameters, should be used in future to prioritize research and development objectives.

The near-term models (A and B) are based on modest extrapolations of the relatively conservative design plasma performance of ITER. Models C and D assume progressive improvements in performance, especially in plasma shaping, stability, and protection of the 'divertor', through which helium 'ash' and impurities will be exhausted. Likewise, while Model A is based on a conservative choice of materials, Models B–D would use increasingly advanced materials and operate at increasingly higher temperatures (which would improve the 'thermodynamic efficiency' with which they turn fusion power into electricity).

The power plant study shows that the cost of fusion generated electricity decreases with the electrical power output (P_e) approximately as $P_e^{-0.4}$. It was assumed that the maximum output acceptable to the grid would be 1.5 GW. Given the increase of temperature and hence thermodynamic efficiency, the size and gross fusion power needed to produce $P_e = 1.5$ GW decreases from model A (with fusion power 5.0 GW) to D (fusion power 2.5 GW). The cost of electricity, which is dominated by the capital cost, also decreases with size from 9 Eurocents/kWhr for an early model A to 5 Eurocents/kWhr for an early model D (these costs would decrease as the technology matures). Even the first cost would be competitive with other generating costs if there was a significant carbon tax, which now effectively exists in Europe with the Emissions Trading Scheme. If acceptable and necessary, larger plants (with $P_e > 1.5$ GW) would be more cost effective.

The power plant study shows that economically acceptable fusion power stations, with major safety and environmental advantages, seem to be accessible through ITER with material testing, if possible in parallel, at IFMIF (but without major material advances).

The above discussion assumes that the first prototype fusion power station will be based on a conventional (ITER/JET-like) tokamak. This will almost certainly be the case, unless ITER produces major surprises. Alternative magnetic confinement configurations ('spherical tokamaks'; 'stellarators') are, however, under development that have certain theoretical advantages, and could form the basis for later prototypes and actual fusion power stations. Meanwhile they provide additional insights into the behaviour of plasmas and fusion technology which feed into the mainstream, conventional tokamak, line.

Fast track studies

A detailed study of the time that will be needed to develop fusion has recently been completed at UKAEA Culham. It was assumed that construction of ITER and IFMIF both begin in the immediate future. The information that will be needed to finalize the design of the first prototype fusion power station, which has become known as DEMO (for Demonstrator), was identified and estimates were made of when this information will be provided by ITER and IFMIF. Assuming just in time provision of the necessary information, this led to the construction timetable for DEMO shown in Fig. 7.7. Commercial fusion power stations would follow some ten or more years after DEMO comes into operation.

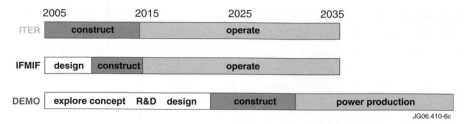

Figure 7.7   Possible time-line for the construction of the devices (ITER and IFMIF) that are needed before a prototype power station (DEMO) can be built, and for construction of DEMO itself. This is a very much simplified summary of the results of a detailed study.

The Culham fast track timetable reflects an orderly, relatively low risk, approach. It could be speeded up if greater financial risks were taken, for example, by starting DEMO construction before in situ tritium generation and recovery have been demonstrated. The risks could be reduced—and the timetable perhaps speeded up—by the parallel construction of multiple machines at each stage.

Fig. 7.7 assumes that ITER and IFMIF are approved at the same time, which is highly desirable but may not be realistic (some delay in IFMIF construction might however be tolerable without comprising the end date). It should be stressed generally that the fast track model is a technically feasible plan, *not* a prediction. Meeting the timetable will require a change of focus in the fusion community to a project orientated 'industrial', approach, accompanied of course by the necessary funding and political backing.

Conclusions

The world faces an enormous energy challenge, as a result of rising energy use, and the fact that burning fossil fuels (which currently provide 80% of primary energy) is driving potentially catastrophic climate change and, when not managed carefully, producing debilitating pollution. The response must be a cocktail of measures: we must strive to use energy more efficiently, and renewables should play a role where appropriate. But there are in principle only four ways of meeting a large fraction of world energy demand: continuing use of fossil fuels (as long as they last); solar power (but realizing its potential requires major breakthroughs); nuclear fission; and fusion.

With so few horses in the race, we cannot afford *not* to back fusion.

Given the remarkable progress that has been achieved in recent decades, we are confident that fusion will be used as a commercial power source in the long term. We are less confident that fusion will be available commercially on the time scale outlined above, which would require adequate funding of a properly focused and managed programme, and that there are no major surprises. However, given the magnitude of the energy challenge and the relatively small investment that is needed on the ($3 trillion pa) scale of the energy market, we are absolutely convinced that accelerated development of fusion would be fully justified.

Acknowledgement

This work was funded by the UK Engineering and Physical Sciences Research Council and Euratom.

Resources and further information

General information on fusion can be found in the recent book, *Fusion, the Energy of the Universe*, G. McCracken and P. Stott, Elsevier (2005), ISBN 0-12-481851-X, and in *La Fusion Nucléaire*, J. Weisse, Presse Universitairs de France (2003), ISBN 2-13-053309-4 or on the UKAEA Culham web site www.fusion.org.uk

The EU studies of fusion power plant concepts are available at http://www.wfda. org/portal/downloads/efda_rep.htm: for a summary see *A Conceptual Study of Commercial Fusion Power Plants*, D. Maisonnier *et al.*, EFDA-RP-RE-5.0 (2004). The analysis of the development path for fusion is given in *Accelerated Development of Fusion Power* by I. Cook *et al.*, February 2005, UKAEA FUS 521, available at http://www.fusion.org.uk/techdocs/ukaea-fus-521.pdf

The authors

Prof Sir Chris Llewellyn Smith FRS is Director of UKAEA Culham Division, which operates JET on behalf of Euratom and houses the UK's fusion research programme. He spent the first part of his career as a theoretical high energy physicist, before becoming Chairman of Oxford Physics, Director General of CERN (1994–98), and President and Provost of University College London. His scientific contributions and leadership have been recognized by awards and honours in seven countries in three continents. He has also spoken widely and written on science policy and energy issues.

Dr David Ward is a fusion scientist at UKAEA Culham. Having first worked on theory and experiments at JET, he now plays a leading role in pan-European fusion power station design studies and studies of the socio-economics of fusion.

8. *Photovoltaic and photoelectrochemical conversion of solar energy*

Michael Grätzel

Introduction

The Sun provides about 100,000 Terawatts (TW) to the Earth, which is approximately ten thousand times greater than the world's present rate of energy consumption (14 TW). Photovoltaic (PV) cells are being used increasingly to tap into this huge resource and will play a key role in future sustainable energy systems. Indeed, our present needs could be met by covering 0.5% of the Earth's surface with PV installations that achieve a conversion efficiency of 10%.

Principles

Fig. 8.1 shows a simple diagram of how a conventional photovoltaic device works. The top and bottom layers are made of an n-doped and p-doped silicon,

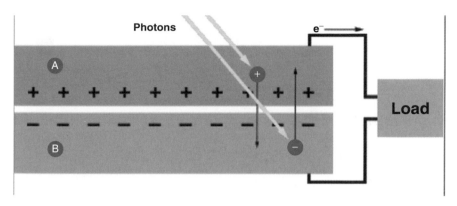

Figure 8.1 Schematic diagram of how a conventional solid state photovoltaic device works.

where the charge of the mobile carriers is negative (electrons) or positive (holes), respectively. The p-doped silicon is made by 'doping' traces of an electron-poor element such as gallium into pure silicon, whereas n-doped silicon is made by doping with an electron-rich element such as phosphorus. When the two materials contact each other spontaneous electron and hole transfer across the junction produces an excess positive charge on the side of the n-doped silicon (A) and an excess negative charge on the opposite p-doped (B) side. The resulting electric field plays a vital role in the photovoltaic energy conversion process. Absorption of sunlight generates electron-hole pairs by promoting electrons from the valence band to the conduction band of the silicon.

Electrons are minority carriers in the p-type silicon while holes are minority carriers in the n-type material. Their lifetime is very short as they recombine within microseconds with the oppositely charged majority carriers. The electric field helps to collect the photo-induced carriers because it attracts the minority carriers across the junction as indicated by the arrows in Fig. 8.1, generating a net photocurrent. As there is no photocurrent flowing in the absence of a field, the maximum photo-voltage that can be attained by the device equals the potential difference that is set up in the dark at the p-n junction. For silicon this is about 0.7 V.

So far, solid-state junction devices based on crystalline or amorphous silicon (Si) have dominated photovoltaic solar energy converters, with 94% of the market share. These systems have benefited from the experience and material availability generated by the semiconductor industry and they are at a mature state of technical development in a rapidly growing market. Fig. 8.2 shows the photovoltaic peak-power installed annually from 1988 to 2003. In 2004 and 2005 the trend has continued, the peak-powers being 1.15 GW and 1.5 GW respectively.

By 2010 the module output sales are expected to quadruple again, reaching 6 GW. This impressive growth is being fuelled by attractive feed-in tariffs.

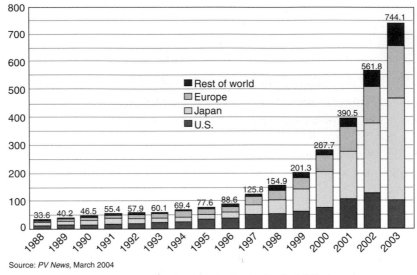

Source: *PV News*, March 2004

World PV Cell/Module Production (in MW)

Figure 8.2   The increase in worldwide production of PV modules during the last 15 years. The market growth rate in the last decade has been over 30%.

Share prices in solar cell companies are soaring and revenues are predicted to increase from 8.3 billion US$ in 2004, to 36.1 billion US$ in 2010, with pre-tax profits rising from 1.2 to 6.4 billion US$ over the same period. By 2030 the yearly PV module output is expected to attain 300 GW (PV-TRAC, 2005). However, this is still too little to make a major impact because the world's energy consumption will by then approach 20 TW; in addition photovoltaic power figures are expressed in peak-watts, that is the output reached only in full sunshine at 1 kW/m^2 incident light intensity. The real power, averaged over day and night and the 4 seasons is about 3–10 times lower (depending on geographical location as well as direction and angle of orientation of the panels). Nevertheless, based on the recent growth rates, it has been predicted (Zweibel, 2005) that by 2065 all the world's energy needs could in principle be met by photovoltaic cells.

Conversion efficiencies

The conversion efficiency of a solar cell is defined as the ratio of its electric power output to the incoming light intensity that strikes the cell. The internationally accepted standard test condition (STC) uses as a reference sunlight (viewed at 48 degrees to overhead) normalized to 1 kW/m^2, with the temperature kept at 298 K. The conversion efficiency is determined by measuring the photocurrent (that is, the electrical current induced by light) as a function of the cell voltage using the formula

$$\eta = J_{sc} V_{oc} FF / I_s \tag{8.1}$$

Here J_{sc} is the short-circuit photocurrent density, V_{oc} the open circuit voltage, I_s the incident solar intensity $(1000\,W/m^2)$ and FF the fill factor defined as the electric power produced at the maximum power point of the J-V curve divided by the product $J_{sc} \times V_{oc}$.

Fig. 8.3 shows the history of confirmed 'champion' laboratory cell efficiencies. The performance of conventional solar cells is approaching a plateau; only incremental improvements have been accomplished in the last decade despite dedicated R&D effort. The efficiency of multi-junction cells based on III/IV semiconductors has progressed recently beyond 30%. However, the cost of these devices is very high, limiting their application to space and solar concentrators. In the latter case sunlight is concentrated typically several hundred times by a mirror or a lens before striking the photovoltaic device. This will reduce cost if the price per square metre of the solar concentrator is below that of the photovoltaic cell. However, solar concentrators need to track the sun and work well only in direct sunlight in the absence of haze, considerably limiting their potential for practical applications.

The efficiencies reached with commercial solar cell modules are significantly lower than those of the best laboratory cells due to losses incurred

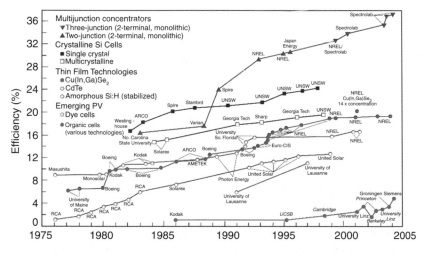

Figure 8.3 A history of best cell efficiencies. All plotted values have been confirmed and were measured under standard test conditions except the multijunction concentrator cells. Source A.J. Nozik , 2004, NREL, USA.

during scale up. The typical size of 'champion laboratory cells' is in the square centimeter range or even below, facilitating the collection of photocurrent. All efforts are made to minimize resistive and optical losses increasing the power output although the methods used may not be applicable or are too expensive for production. Also, commercial criteria influence the choice of methods and materials used for large-scale production and compromises are often made in order to cut costs. Finally large-scale modules need current collector grids or interconnects between individual cells, which reduce the exposed photoactive area of the cell, decreasing the module conversion efficiency.

Conversion efficiencies of solar cells under real conditions can differ significantly from the values measured at STC. One reason for this discrepancy is that in full sunlight the temperature of a module rises within minutes to over 60°C. Because the efficiency of silicon solar cells drops by 0.5% per °C a module with a specified efficiency of say 12% at STC would exhibit less than 10% efficiency in real sunshine. The temperature dependence of efficiency means that a cell located in a cloudy, temperate zone could actually be more efficient than one in a sunny desert, although the absolute power produced would of course be lower than in direct sunshine.

The advantage of the new dye-sensitized solar cells discussed further below is that their output remains constant between 25 and 65°C. They are also less sensitive to the angle of incidence in capturing solar radiation than the conventional silicon cells. Consequently, at equal STC rating they fare 20–40% better than conventional devices in harvesting solar energy under real outdoor conditions.

Cost and supplies of raw materials

While the growth of the PV market over the last decade has been impressive, the cost of photovoltaic electricity production is still too high to be competitive with nuclear or fossil fuel. For the best systems installed at well-chosen sites, the price per kWh is at present 0.25–0.65 US\$. In order to be competitive, the price would have to come down to below 0.05 US\$/kWh. This ambitious goal implies that the total cost of the installed PV system should decrease below 1 US\$/Watt. The module itself should contribute less than 0.5 US\$/Watt to this price, a target that seems difficult if not impossible to meet with present silicon technology.

A major dilemma that the PV industry currently faces is a shortage of raw materials. The silicon for today's PV cells originates primarily from waste produced by the chip industry. Alarmingly, the cost of solar-grade silicon leapt from US\$9 per kilo in 2000, to US\$25 in 2004, and US\$60 in 2005. Because 13 grams of silicon are needed to generate one Watt of electric power in full sunlight, i.e. one 'peakwatt', the cost contribution from the raw materials alone

Table 8.1 Examples of three generations of photovoltaic cells

1st Generation	*2nd Generation (low cost, mainly thin films)*
Single crystal	Amorphous Si
Poly-crystalline (silicon)	Thin-film Si
	CuIn(Ga)Se$_2$, CdTe
	Dye-sensitized nanocrystalline Cells (DSC)
	Organic PV (molecular and polymeric)

3rdGeneration (maximum conversion efficiency above the 33% limit for single junction converters in AM 1.5 sunlight).

Multi-gap tandem cells
Hot electron converters
Carrier Multiplication cells
Mid-band PV
Quantum Dot Solar Cells

is already 0.78 US\$/Watt, rendering it difficult to meet the 0.5 US\$/Watt target unless the silicon price falls substantially or the quantity required to produce a peakwatt can be greatly reduced. The higher prices result from fast growth of the global PV industry and the increased efficiencies in the traditional silicon chip industry which results in less waste available for PV production. Some of the world's major solar-cell makers are warning of a 'vicious spiral' in which the market would grind to a halt as silicon prices rocket and supplies do not meet demand, severely curbing the annual growth rates of 30 to 40% that the industry has witnessed since 1997. Thus for 2006, growth is expected to be only 5%. In the near to medium term, a rise in module cost by at least 15% from the current value of 3.7 US\$/Watt to 4.50 US\$/Watt seems inevitable (Rogol, 2005). Clearly, a change in economics is required for photovoltaics to become fully cost competitive.

How PV cells are developing

Second generation thin film PV cells

Since the 1970s new generations of PV cells have emerged and the main examples are listed in Table 8.1. The traditional thin-film photovoltaic cells made of CuInSe (copper indium selenide) or CdTe (cadmium telluride), along with amorphous silicon, have been around for several decades and are being increasingly employed. Their market share is expected to grow significantly from the current 5%. The conversion yields of commercial devices are still significantly below the 12–17% attained by polycrystalline and single-crystal silicon but the energy pay-back time is shorter. For silicon and conventional thin-film cells the pay-back time is 3–4 years, see Fig. 8.4. Pay-back times for the dye-sensitized and organic thin film PVs are expected to be below one year.

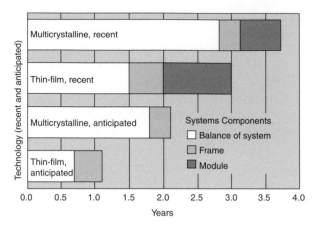

Figure 8.4   Energy pay-back times for silicon and thin-film PV cells (Source: NREL, 2004).

Although CuIn(Ga)Se and CdTe devices are attractive due to their high conversion efficiency, which reaches between 15 and 20% for best laboratory demonstrations, they are unlikely to become large-scale suppliers of solar electricity due to the scarcity of indium, tellurium, and selenium, and the high toxicity of cadmium (Green, forthcoming). The prices of these elements have decoupled recently and are now 1000, 180, and 150 US$/kg for In, Te, and Se, respectively. Although much smaller material quantities are needed for thin-film PV devices, i.e. about 100 mg/Watt (Keshner and Arya, 2004) compared to 13 g/Watt for silicon, low availability and environmental concerns remain a problem.

Most of the annual 50–70 MW thin-film market is currently served by amorphous silicon (a-Si) solar cells. These give lower efficiencies, i.e. 5–7%, instead of 7–11% obtained with CdTe- and CuInSe-based devices and are subject to performance degradation. The need for high vacuum production methods makes them as expensive on a $/peakwatt basis as crystalline silicon cells.

Mesoscopic solar cells

These new solar cells employ films composed of a network of inorganic or organic semiconductors particles of mesoscopic (2–50 nm) size, forming junctions of very high contact area instead of the flat morphology used by conventional thin-film cells. They are commonly referred to as 'bulk' or 'interpenetrating network' junction cells due to their three-dimensional structure. The prototype of this family of devices is the dye-sensitized solar cell (DSC), which accomplishes optical absorption and charge separation by combining a light-absorbing material (the sensitizer) with a wide band-gap semiconductor of mesoscopic morphology (O'Regan and Gratzel, 1991). The DSC is used in conjunction with electrolytes

(Gratzel, 2001), ionic liquids (Wang *et al.*, 2005), polymer electrolytes (Haque *et al.*, 2003), or organic (Bach *et al.*, 1998) as well as inorganic hole conductors (Perera *et al.*, 2003; O'Regan *et al.*, 1007). Other strategies employ blends of organic materials, such as polymer (Halls *et al.*, 1995) or molecular semiconductors (Peumans and Forrest, 2001) as well as hybrid cells using a p-type semiconducting polymer (such as poly-3-hexylthiophene), in conjunction with a fullerene (Brabec *et al.*, 2003) or CdSe 'nanorods' (Huynh *et al.*, 2002).

These new dye-sensitized solar cells could be fabricated without expensive and energy-intensive high temperature and high vacuum processes. They are compatible with various supporting materials and can be produced in a variety of presentations and appearances to enter markets for domestic devices and architectural or decorative applications. The presence of a mesoscopic junction having a large-area interface confers to these devices intriguing optoelectronic properties. The DSC has shown conversion efficiencies exceeding 11%, rivaling conventional devices. Excellent stability under long-term illumination and high temperatures has been reached, attracting industrial applications. Photo-electrochemical cells also offer a way to generate hydrogen via solar photolysis using, for example, tandem cells. This offers a way to convert solar energy directly into a chemical fuel (Gratzel, 2001).

Fig. 8.5, shows band diagrams for a conventional p-n junction photovoltaic device (left side) and a dye-sensitized solar cell (right side). In conventional p-n junction solar cells the materials used must be very pure to obtain a good conversion yield and should not contain lattice imperfections or grain boundaries which accelerate electron-hole recombination. A critical device parameter is the diffusion length of the minority charge carriers, i.e. conduction-band electrons and valence-band holes for the n-doped and p-doped material, respectively. The diffusion length should be greater than the thickness of the film in order to collect all the charge carriers generated by light excitation.

By contrast mesoscopic injection solar cells operate in an entirely different fashion from conventional solar p-n junction devices. Mimicking the principles that natural photosynthesis has used successfully over the last 3.5 billion years in biological solar energy conversion, they achieve the separation of light-harvesting and charge-carrier transport. The semiconductors used in conventional cells assume both functions simultaneously, imposing stringent demands on purity and entailing high materials and production cost. The prototype of this new PV family is the dye-sensitized solar cell (DSC) whose principle of operation is shown on the right side of Fig. 8.5.

The benefits of the mesostructure

The mesoscopic morphology of materials used in these new thin film PV devices is essential for their efficient operation. For the DSC the nanocrystalline structure of

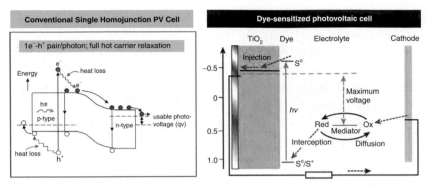

Figure 8.5 Band diagrams for a conventional photovoltaic device (left) and a dye-sensitized solar cell (right). The band bending at the p-n junction is produced by the spontaneous dark current flow depicted in Fig. 8.1. Absorption of a photon produces a single electron-hole pair (charge separation) and some heat. Note that the semiconductor on the left assumes two functions simultaneously: it absorbs sunlight and transports electric current. In order to collect all the charge carriers generated by light excitation, the materials must be very pure and should not contain lattice imperfections or grain boundaries accelerating electron-hole recombination. A critical device parameter is the diffusion length of the minority charge carriers (conduction band electrons for the p-doped Si and valence band holes for the n-doped side) which should be larger than the thickness of the film. By contrast, in the dye-sensitized cell (DSC), the light absorption and charge carrier transport are separated and there are no minority charge carriers involved in the photo-conversion process. Light is absorbed by a dye molecule attached to the surface of a nanocrystalline titanium dioxide film, which is selected because it is stable and environmentally friendly. During operation, the sensitizer transfers an electron to the semiconductor. The dye molecule, now positively charged, is regenerated by the electrolyte or a p-type hole conductor while the latter recovers an electron from the external circuit. The conversion efficiency of the DSC is currently above 11% while silicon champion cells have attained 24%, the maximum being 33% for both devices. The use of nonvolatile or solvent-free electrolytes in the DSC gives very stable performance.

the oxide semiconductor used to support the sensitizer has the following benefits:

1. It renders possible efficient light harvesting by the surface absorbed sensitizer (Gur *et al.*, 2005). On a flat surface a monolayer of dye absorbs at most a few percent of light because it occupies an area that is several hundred times larger than its optical cross section. Using multi-layers of sensitizer does not offer a viable solution to this problem because only those molecules that are in direct contact with the oxide surface are photoactive, the rest act as a filter attenuating the light that strikes the sensitizer. The huge amplification of the interfacial area enhances the light absorption resulting in a 1,000-fold increase in the photocurrent compared to a DSC having a flat surface morphology.

2. The TiO_2 nanocrystals do not have to be electronically doped to render them conducting because the injection of one electron from the sensitizer into a 20-nm sized TiO_2 particle suffices to switch the latter from an insulating to a

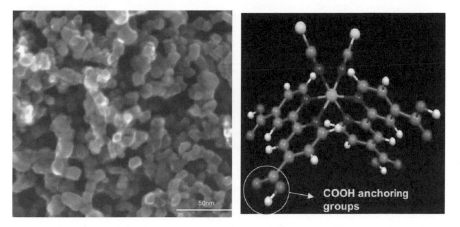

Figure 8.6   *Left*: scanning electron microscopy picture of a nanocrystalline TiO$_2$ film used in the DSC. *right*: molecular structure of a compound (cis-Ru(2,2′-bipyridyl-4,4′-dicarboxlyate)$_2$ (SCN)$_2$) that is widely used as a sensitizer.

conductive state. This photo-induced conductivity of the particle films allows the electrons to be collected without significant ohmic loss. By contrast, a compact semiconductor film would need to be n-doped to conduct electrons. In this case, energy transfer from the excited sensitizer to the conduction band electrons of the semiconductor would inevitably reduce the photovoltaic conversion efficiency.

3. The small dimension of the TiO$_2$ particles allows for efficient screening of the negative charge of the electrons by the electrolyte or hole conductor present in the pores. As a result the photocurrent is not impaired by the repulsive interactions between electrons diffusing through the particle network.

Fig. 8.6 shows a scanning electron microscopy image of a mesoscopic TiO$_2$ (anatase) layer and the molecular structure of the most frequently used sensitizer (light harvester). The particles have an average size of 20 nm (20 billionths of a meter). By coating the oxide nanocrystals with a monolayer of sensitizer it is possible to produce far more efficient solar energy conversion devices.

Opportunities for performance improvement

The DSC currently reaches over 11% energy conversion efficiencies under standard reporting conditions in liquid junction devices (Graetzel, 2005), rendering it a credible alternative to conventional p-n junction photovoltaic devices. Typical photovoltaic performance data are shown in Fig. 8.7. Solid-state equivalents using organic hole-conductors have exceeded 4% efficiency (Schmidt-Mende *et al.*, 2005) whereas nanocomposite films comprising only

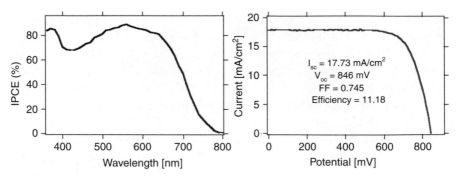

Figure 8.7 Photovoltaic performance of a 'state of the art' DSC laboratory cell. Left: photo-current action spectrum showing the monochromatic incident photon to current conversion efficiency (IPCE) as function of the light wavelength obtained with the N-719 sensitizer. Right: J-V curve of the same cell under AM 1.5 standard test conditions.

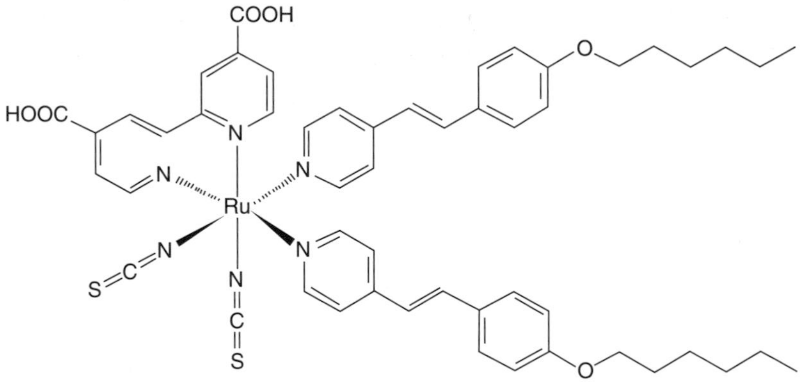

Figure 8.8 Structure of K-19, a new dye which exhibits enhanced light absorption.

inorganic materials, such a TiO_2 and $CuInS_2$ have achieved efficiencies between 5 and 6% (Nanu *et al.*, 2004; 2005) which is significantly higher than the recently reported CdTe/CdSe particle-based heterojunctions (Gur *et al.*, 2005). Organic PV cells based on blends of a fullerene derivative with poly-3-hexylthiophene have a confirmed conversion efficiency of 4.8%. (Brabec *et al.*, 2003).

To further improve the DSC performance, new dyes showing increased optical cross sections, and capable of absorbing longer wavelengths, are under development. Similarly, ordered mesoscopic TiO_2 films have been introduced as current collectors, benefiting greatly from recent advances in nano-material research (Zukalova *et al.*, 2005). Interfacial engineering has produced major gains in the open circuit voltage (Zhang *et al.*, 2005). A new generation of sensitizers is being developed, having higher optical cross sections (Wand *et al.*, 2005) and enhanced near-IR response. Shown in Fig 8.8 is the structure of a new dye (codename K-19) which exhibits enhanced light absorption.

The new IR sensitive dyes or dye cocktails are expected to reach photocurrent densities of 25–27 mA/cm^2, which should allow the overall conversion efficiency of the DSC to exceed 15%, without changing the currently used non-volatile redox electrolyte. A roadmap to achieve this goal within the next two years has been established and will serve to coordinate synthetic efforts of several international groups. Importantly, DSCs based on the K-19 sensitizer have shown excellent stability both under long term light soaking and thermal stress (Wang *et al.*, 2005).

Third generation photovoltaic cells

Research is also booming in the area of so-called 'third generation' photovoltaic cells. A recent breakthrough concerning charge carrier multiplication offers great promise. Enormous excitement has followed the prediction, (Nozik, 2004) discovery, (Schaller and Klimov, 2004) and confirmation (Nozik 2005; Hanna & Nozik 2006) that several excitons can be produced from the absorption of a single photon by very small semiconductor particles, called 'quantum dots' because their electronic properties are different from those of bulk-size materials due to the confinement of the electron-hole pairs produced by optical excitation. This effect occurs via impact ionization (IMI) if the photon energy is several times higher than the semiconductor band gap. The challenge is now to find ways to collect the excitons before they recombine. As recombination occurs on a picosecond time-scale, the use of mesoporous oxide collector electrodes presents a promising strategy, because transfer of the electron from the quantum dot to the conduction band of the oxide collector electrode can occur within femtoseconds (Nozik, 2004; Plass *et al.*, 2002). This opens up research avenues that ultimately may lead to photo-converters reaching external quantum efficiencies of several hundred percent. The reason for this is that optical excitation of a semiconductor quantum dot such as PbSe by a light quantum that has an energy several times higher than the band gap can produce several electron-hole pairs instead of a single pair for a bulk semiconductor as shown in Fig. 8.5. A calculation shows that the maximum conversion efficiency of a single junction cell could be increased from 34% to 44% by exploiting IMI effects (Nozik 2005; Hanna & Nozik 2005).

Field tests of DSC modules

At STC, the conversion efficiency of DSC modules is still about a factor of two below that of commercial silicon modules. However the efficiency gap narrows significantly when the cells are compared under real outdoor conditions as shown by recent field tests of DSC modules performed by Aisin Seiki under realistic outdoor conditions have revealed significant advantages compared to

Figure 8.9   Outdoor field tests of DSC modules produced by Aisin Seiki in Kariya City. Note the pc-Si modules in the second row (photograph courtesy of Aisin Seiki).

silicon panels. Thus, for modules with equal rating under STC, the DSC modules produced 20–40% more energy than the poly-crystalline silicon (pc-Si) modules.[1] A photograph of a test station comparing the two types of PV technologies is shown in Fig. 8.9.

The superior performance of the DSC can be ascribed to the following facts:

- The DSC efficiency is practically independent of temperature in the range 25 to 65°C, whereas that of pc-Si declines by $\sim 20\%$ over the same range.

- The DSC collects more solar energy during the day than pc-Si due to a lower sensitivity of the light harvesting to the angle of incidence.

- The DSC shows higher conversion efficiency than pc-Si in diffuse light or cloudy conditions.

Although it is up to the commercial supplier to set the price of completed modules, it is clear that the DSC shares the cost advantage of all thin-film devices. In addition it uses only cheap and readily available materials (Toyoda 2006) and in contrast to amorphous silicon, avoids cost-intensive high vacuum production

---

[1] It may be argued that the ruthenium based sensitizer adds high material cost. However the contribution is less than 0.01€/pWatt given the small amount employed. Also purely organic sensitizers have reached practically the same yield as ruthenium complexes.

Figure 8.10 Small scale flexible 'solar leaf' (USA) and large-scale solar roof (Australia).

steps. Given these additional advantages at comparable conversion efficiencies, module costs below 1 € are realistic targets for large-scale production plants.

Industrial DSC development

The potential of the dye-sensitized solar cell as a practical device is supported by commercial interest, with several industries taking patents and using prototype applications. The DSC panels shown in Fig. 8.9 have been installed in the walls of the Toyota dream house offering an integrated source of solar power to the inhabitants. American and Japanese development engineers are exploiting the potential for incorporation into flexible polymer base materials (www.konarka.com) on which the semiconductor is deposited as a thin film. There are also projects in Australia (www.dyesol.com) and Japan to erect and evaluate large-area systems. Fig. 8.10 shows examples of such applications. Recently a major breakthrough in the commercialization of the DSC has been achieved with the establishment of a 20 MW production plant by G24 Innovations in Wales (www.g24i.com).

Summary

Whereas the present photovoltaic market is dominated by single crystal and polycrystalline silicon cells, we will be looking to the second and third generation cells to contribute the lion's share of the future TeraWatt-scale output in solar module production, which is required to make a major contribution towards the huge future demands for renewable energy. Mesoscopic cells are well suited for a whole realm of applications, ranging from the low-power market to viable large-scale applications. Thus, their excellent performance in diffuse light gives them a competitive edge over silicon in providing electric power for stand-alone electronic equipment, both indoors and outdoors. Application of the DSC in building integrated PV cells has already started and will become a fertile field of future commercial development. With the ongoing expansion of the PV market and with the escalation of fossil fuel prices as well as environmental considerations, there is high expectation that mesoscopic cells will play a significant role in providing solar power in competition with conventional devices and other innovations.

Acknowledgements

Financial support from EU and Swiss (ENK6-CT2001-575 and SES6-CT-2003-502783), and USAF (contract No. FA8655-03-13068) is acknowledged

Resources and further information

Bach, U., Lupo, D., Comte, P., Moser, J. E., Weissoertel, F., Salbeck, J. Spreitzert, H., and Graetzel, M. 1998. Solid state dye sensitized cell showing high photon to current conversion efficiencies. *Nature*, **395**, 550.

Brabec, C. J., Dyakonov, V., Parisi, J., and Sariciftci, N. S. (Eds) 2003, Organic Photovoltaics: Concepts and Realization. *In: Springer Ser. Mater. Sci.,* **6**.

Chiba, Y., Islam, A., Watanabe, Y, Komiya, R., Koide, N., and Han, L. 2006. Dye-sensitized solar cells with conversion efficiency of 11.1% Jap.J.Appl.Phys. part 2: Letters & Express Letters **45**, 24–8.

European Photovoltaic Technology Research Advisory Council (PV-TRAC).2005. *A Vision for Photovoltaic Energy Production*, EUR 21242, Office for Official Publications of the European Communities, Luxembourg.

Graetzel, M. 2001. Photoelectrochemical cells. *Nature*, **414**, 338–44.

Graetzel, M. 2005a. Charge separation and efficient light energy conversion in sensitized mesoscopic photoelectrochemical cells based on binary ionic liquids. *J. Am. Chem. Soc.* **127**, 6850–56.

Graetzel, M. 2005b. Mesoscopic solar cells for electricity and hydrogen production from sunlight. *Chem. Lett.* **34**, 8–13.

Green, M.A. forthcoming. Consolidation of Thin-Film Photovoltaic Technology, the Coming decade of Opportunity, Progr. Photovolt. Eng

Gur, I., Fromer, N. A., Geier, M. L., and Alivisatos, A. P. 2005. Air-stable all-inorganis nanocrystal solar cells processed from solution, Science **310**, 462–65.

Halls, J. J. M., Walsh, C. A., Greenham, N. C., Marseglia, E. A., Friend, R. H., Moratti, S. C., Holmes, A. B. Efficient photodiodes from interpenetrating polymer networks. *Nature* **376**, 498–500 (1995).

Hanna, M. C. and Nozik, A. J. 2006. Solar conversion efficiency of photovoltaic and photoelectrolysis cells with carrier multiplication absorbers. J. Appl.Phys, **100**, 074510/1-074510/8.

Haque, S. A., Palomares, E., Upadhyaya, H. M., Otley, L., Potter, R. J., Holmes, A. B., and Durrant, J. R. 2003. Flexible dye sensitised nanocrystalline semiconductor solar cells. *Chem. Comm.* **24**, 3008–09.

Henry, C. H. 1980. Limiting efficiencies of ideal single and multiple energy gap terrestrial solar cells, *J. Appl. Phys.* **51**, 4494–4500.

Huynh, W. U., Dittmer, J., and Alivisatos, A. P. 2002. Hybrid nanorod-polymer solar cells, *Science* **295**, 2425–2427.

Keshner, M. S. and Arya, R. 2004. *Study of potential cost reductions resulting from super-largescale manufacturing of PV modules.* Final Subcontract Report, NREL/SR-520-56846, October, 2004.

Nanu, M., Schoonman, J., and Goossens, A., 2004. Inorganic nanocomposites of n- and p-type semicinductors: a new type of three-dimensional solar cell, *Advanced Materials* **16**, 453–456.

Nanu, M., Schoonman, J. and Goossens, A. 2005. Solar energy conversion in $TiO_2/CuInS_2$ nanocomposites, *Adv. Func. Mat.* **15**, 95–100.

Nozik, A. J. 2004. Quantum dot solar cells, *Next Generation Photovoltaics*, 196–222.

Nozik, A. J. 2005. Exciton multiplication and relaxation dynamics in quantum dots: applications to ultrahigh-efficiency solar photon conversion. *Inorg. Chem.* **44**, 6893–6899.

O'Regan, B. and Graetzel, M. 1991. cw cost and highly efficient solar cells based on tbe sensitization of colloidal titanium dioxide, Nature **335**, 7377.

O'Regan, B., and Schwartz, D. T. 1997. Solid state photoelectrochemical cells based on dye sensitization. *AIP Conf. Proc.* **404** (Future Gen. Photovolt. Techn.), 129–36.

Perera, V. P. S., Pitigala, P. K. D. D. P., Jayaweera, P. V. V., Bandaranayake, K. M. P., and Tennakone, K. 2003. Dye-sensitized solid-state photovoltaic cells based on dye multilayer-semiconductor nanostructures. *J. Phys. Chem. B* **107**, 13758–61 (2003).

Peumans, P., Forrest, S. R. 2001. Very-high-efficiency double-heterostructure copper phthalocyanine/C60 photovoltaic cells. Appl. Phys. Lett. **79**, 126–8.

Plass, R., Pelet, S., Krueger, J., Graetzel, M., and Bach.U. 2002. J. Phys. Chem. B **106**, 7578–80.

The National Renewable Energy Laboratory, 2004, http://www.nrel.gov/docs/fy04osti/35489.pdf

Rogol, M. 2005a. *Sunscreen II, Investment Opportunities in Solar Power*, CSLA Report, July 2005.

Rogol, M. 2005b. http://www.photononsulting.com/executive_summary.htm

Schaller R. and Klimov, V. I. 2004. High efficiency carrier multiplication in PbSe nanocrystals: implications for solar energy conversion, Phys. Rev. Lett. **92**, 186601–604.

Schmidt-Mende, L., Bach, U., Humphry-Baker, R., Horiuchi, T., Miura, H., Ito, S., Uchida, S., and Graetzel, M. 2005. Organic dye for highly efficient solid-state dye-sensitized solar cells. Adv. Mater. **17**, 813–15

Toyoda, T. 2006. (Aisin Seiki, Japan) Realization of large area DSC modules and their outdoor performance, *DSC symposium, International RE2006 Congress*, Chiba Japan. http://www.toyota.co.jp/jp/news/04/Dec/nt04_1204.html.

Wang, P., Klein, C., Humphry-Baker, R., Zakeeruddin, S.M., and Graetzel.M. 2005a, Thermal stability of ion gels. J. Am. Chem Soc. **127**, 808.

Wang. P., Klein, C., Humphry-Baker, R., Zakeeruddin, S. M., and. Graetzel, M. 2005b. Efficient nanocrystalline dye-sensitized solar cell based on an electrolyte of low volatility, *Appl. Phys. Lett.* **86**, 123508

Wang, P., Wenger,B., Humphry-Baker, R., Moser, J.-E. Teuscher, J., Kantlehner, W., Mezger, J. ,Stoyanov, E. V., Zakeeruddin, S. M., and Graetzel, M, 2005, Charge separation and efficiemt light energy conversion in sensitized mesoscopic photoelectrochemical cells based on binary ionic liquids, *J, Am. Chem. Soc.* **127**, 6850–56.

Zukalova, M., Zukal, A., Kavan, L., Nazeeruddin, M. K., Liska, P., and Graetzel, M. 2005. Organized mesoporous TiO2 films exhibiting greatly enhanced performance in dye-sensitized solar cells. *Nano Lett.* **5**, 1789–92.

Zhang, Zh,. Zakeeruddin, S. M., O'Regan, B. C., Humphry-Baker, R., and Graetzel, M. 2005. Influence of 4-Guanidinobutyric acid as coadsorbent in reducing recombination in dye-sensitized solar cells. J. Phys. Chem. B **109**, 21818–24.

Zweibel, K. 2005. The Terawatt Challenge for Thin-film PV, Technical Report NREL/TP-52038350.

The author

Michael Gratzel is Professor at the Ecole Polytechnique Fédérale de Lausanne, where he directs the Laboratory of Photonics and Interfaces. He discovered a new type of solar cell based on dye sensitized mesoscopic oxide particles and pioneered the use of nanomaterials in electroluminscent and electrochromic displays as well as lithium ion batteries. Author of over 500 publications, two books and inventor of over 50 patents, he ranks amongst the most highly cited scientists in the world. He has received numerous prestigious awards including the 2006 World Technology award, the Gerischer medal of the Electrochemical Society, the Italgas prize, the Faraday Medal of the British Royal Society of Chemistry, the 2000 European Millennium innovation award, and the Dutch Havinga award. He received a doctors degree in natural science from the Technical University Berlin and honorary doctors degrees from the Universities of Delft, Uppsala, and Turin. He is a member of the Swiss Chemical Society as well as of the European Academy of Science and was elected honorary member of the Société Vaudoise de Sciences Naturelles.

9. *Biological solar energy*

James Barber

Introduction

Oil, gas, and coal provide us with most of the energy needed to power our technologies, heat our homes, and produce the wide range of chemicals and materials that support everyday life. Ultimately the quantities of fossil fuels available to us today will dwindle, and then what? Even before that we are faced with the problem of increasing levels of carbon dioxide in the atmosphere and the consequences of global warming (Climate Change, 2001). To address these issues it is appropriate to remind ourselves that fossil fuel reserves are derived from the process of photosynthesis. Plants, algae, and certain types of bacteria have learnt how to capture sunlight efficiently and convert it into organic molecules, the building blocks of all living organisms. It is estimated that photosynthesis produces more than 100 billion tons of dry biomass annually, which would be equivalent to a hundred times the weight of the total human population on our planet at the present time, and equal to about 100 TJ of stored energy.

In this chapter we emphasize the enormity of the energy/carbon dioxide problem that we face within the coming decades and discuss the contributions that could be made by biofuels and developing new technologies based on the successful principles of photosynthesis. We will particularly emphasize the possibility of exploiting the vast amounts of solar energy available to extract hydrogen directly from water.

Principles of photosynthesis

The success of this energy generating and storage system stems from the fact that the raw materials and power needed to synthesise biomass are available in almost unlimited amounts: sunlight, water and carbon dioxide. At the heart of the reaction is the splitting of water by sunlight into oxygen and hydrogen. The oxygen (a 'waste product' of the synthesis) is released into the atmosphere where it is available for us to breathe and to use for burning our fuels. The 'hydrogen' is not normally released into the atmosphere as H_2, but instead is combined with carbon dioxide to make organic molecules of various types. When we burn fossil fuels we combine the 'carbon-stored hydrogen' of these organic molecules with oxygen, releasing water and carbon dioxide and effectively reversing the chemical reactions of photosynthesis. Similarly, energy is also released from organic molecules when they are metabolized within our bodies by the process of respiration. It is important to appreciate that all energy derived from the products of photosynthesis (food, biomass, fossil fuels) originates from solar energy (Fig. 9.1).

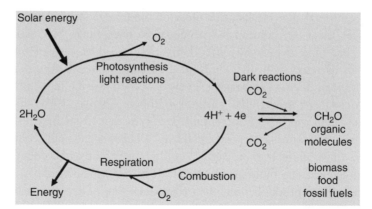

Figure 9.1 Energy flow in biology. The light reactions of photosynthesis (light absorption, charge separation, water splitting, electron/proton transfer) provides the reducing equivalents or 'hydrogen' electrons (e) and protons (H^C) to convert carbon dioxide (CO_2) to sugars and other organic molecules that make up living organisms (biomass) including those that lie at the bottom of food chains and provide food. The same photosynthetic reactions gave rise to the fossil fuels formed millions of years ago. Oxidation of these organic molecules either by respiration (controlled oxidation within our bodies) or by burning fossil fuels is the reverse of photosynthesis, releasing CO_2 and combining the 'hydrogen' back with oxygen to form water. In so doing energy is released, energy which originated from sunlight.

Efficiency of photosynthesis

To estimate the efficiency of photosynthesis several factors must be considered, which we outline with reference to Fig. 9.2.

a) Although photosynthetic organisms can efficiently trap light energy at all wavelengths of visible solar radiation, the energy used for splitting water and 'fixing' carbon dioxide ('fixing' means converting it to carbohydrate) is only equivalent to the red region of the spectrum. Higher energy photons (shorter wavelength light, towards the violet region) are degraded to heat by internal conversion within the light harvesting pigments to the energy level of 'red' photons.

b) For each electron/proton extracted from water and used to reduce CO_2, the energy of two 'red' photons is required. This is accomplished by linking together, in series, two different photosystems: photosystem II (PSII),

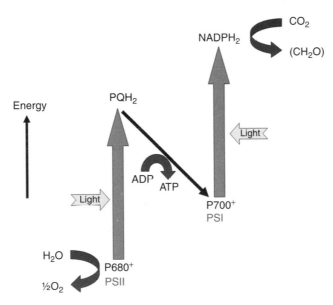

Figure 9.2 Simplified scheme of the light reactions of photosynthesis. Each electron extracted from water and transferred to CO_2 requires the energy of two photons of light. One is absorbed by Photosystem II (PSII) which generates a strong oxidizing species (P680^C), able to drive the water splitting reaction and, a reductant, plastoquinol (PQH$_2$). The other generates a strong reducing species, NADPH$_2$ which donates 'hydrogen' to CO_2 to produce sugars and other organic molecules, symbolized as (CH$_2$O) and a weak oxidant P700^C. Electron and proton flow from PQH$_2$ to P700^C results in the release of energy to convert ADP to ATP. The ATP produced is a store of energy needed, along with NADPH$_2$, to fix CO_2. Since the production of O_2 involves the splitting of two water molecules, the overall process requires eight photons of light.

which uses light to extract electrons/protons from water; and photosystem I (PSI) which uses light to give additional energy to the 'PSII-energized' electrons/protons to drive the CO_2 fixation process (see Fig. 9.2). Therefore photosynthesis uses the energy of at least 8 'red' photons per O_2 molecule released or CO_2 molecule fixed. A typical product of carbon fixation is glucose ($C_6H_{12}O_6$) whose energy content is 2800 kJ per mole (16 kJ/gram). To make a glucose molecule, the energy of 48 'red' photons is required and assuming a wavelength of 680 nm, corresponding to 176 kJ per quantum mole, the efficiency of conversion is 33%. Although this is an impressive number when compared, for example with most photovoltaic devices (see Chapter 8) in reality the overall conversion of solar energy to organic matter is much lower. Energy is lost as higher energy light 'red' photons and is used to drive the enormous number of chemical reactions that occur in photosynthetic organisms to maintain their organization, metabolism, and survival.

There are many ways to define and calculate photosynthetic efficiencies but the approach adopted by Thorndike (1976) is attractive since it engulfs the whole range of definitions. He considered the free energy G stored per photon:

$$G = \eta_T \, \eta_R \, \eta_S \, \eta_L \, \eta_O \, h\nu_O$$

where $h\nu_O$ is the energy of a photon at the optimum frequency for conversion (i.e. red photon), η_O is the thermodynamic efficiency (conversion from energy to free energy produces an explicit entropy T η_S loss), η_L is a factor accounting for irreversible energy losses in photochemical and biochemical pathways, η_S is a factor accounting for the spectral distribution of light and the fact that there is a minimum usable photon energy (close to $h\nu_O$), η_R is a correction factor for leaf reflectivity and η_T is a correction for saturation effects. Taking $h\nu_O$ as 680 nm, and adopting reasonable values for the various coefficients ($\eta_O = 0.73, \eta_L = 0.50, \eta_S = 0.32, \eta_R = 0.80, \eta_T = 0.5$), a maximum efficiency for the conversion of light to stored chemical energy (dry carbon matter) of about 4.5% can be calculated (Thorndyke and Walker,1979, Archer and Barber, 2004, and by others: see Bolton 1977, 1979).

In fact, this value is rarely reached. Only in exceptional cases will the yield of dry matter exceed 1–2%, an exception being the intensive cultivation of sugar cane in tropical climates. Normally agricultural crops yield biomass at efficiencies less than 1%, even when pampered with ample supplies of fertiliser and water. Environmental conditions, degree of light interception, nutrient and water supply are key factors in lowering efficiency, whereas the genetic characteristics of particular plant species also dictates growth rates and maximum biomass.

On a global basis the efficiency of photosynthesis is significantly lower than for optimal agricultural crops, due to seasonal changes and the existence of large areas of land on our planet that do not sustain vegetation. Thus the 4×10^{21} J of energy trapped annually as fixed carbon by photosynthesis represents only about 0.1% conversion of solar energy, given that the solar flux is 100,000 TW. This energy is mainly stored in wood and fibres of terrestrial trees and plants. A similar amount of photosynthetic activity occurs in the oceans but the carbohydrate is rapidly consumed by entering the food chain. Therefore, overall, the efficiency of global photosynthesis is about 0.2%. The fixed carbon provides biomass which was the traditional source of energy for mankind before the exploitation of fossil fuels. It is not surprising, therefore, that there is growing interest in returning to the use of biomass as an alternative to fossil fuels, since its production and use is carbon dioxide neutral.

Biomass

Wood and other forms of biomass can be used to generate heat, electricity, biogas (mainly methane and carbon dioxide), syngas (hydrogen and carbon monoxide), and other biofuels (mainly ethanol). Many organizations consider 'biomass power' as an increasingly attractive option to replace fossil fuels, including the European Union, US Department of Energy (USDOE), and many national government departments and agencies, major companies and utilities in countries like Brazil, Finland, Sweden, UK, and elsewhere. Currently, the global use of biomass is equivalent to about 1.4 TW (see Fig. 9.3). In the US, biomass surpasses hydroelectric power as a source of renewable energy in providing over 3% of the country's energy consumption, corresponding to about 0.1 TW. However, a report from the US Departments of Energy (USDOE) and Agriculture (USDA, 2005) has concluded that biomass could provide the US with about 30% of its present total energy needs. This would be achieved by using non-arable agricultural land and maximizing forestry usage to generate 1.3 billion tons of dry biomass annually, providing about 1 TW power. This projection also relies on plant breeding and genetic engineering strategies to produce new cultivars for high yields of biomass with minimal input of fertilizers, water, and pesticides. Moreover, improved technologies will be required to maximize the use of biomass including those for producing liquid biofuels to replace gasoline. Calculations must take into consideration the energy costs of maintaining 'energy farms', harvesting the biomass and transporting to a central location for use.

The US and many other countries such as Brazil, are blessed with large land masses and good conditions for growing biomass crops and trees. In contrast, a small, industrial country like the UK has very little spare land available to devote to large scale biomass production for energy. Despite this, the current estimate is that biomass contributes 1.4% to the total energy consumption in the UK, although some of this includes burning domestic waste or utilizing methane gas generated

Figure 9.3 Sugar cane, an energy crop yielding biomass with high efficiency of solar energy conversion at about 1%. In Brazil large areas are devoted to growing this crop for ethanol production by fermentation. Ethanol is used to replace or supplement gasoline for automobiles.

at land fill sites. Reports from the UK Carbon Trust (2005) and the UK Biomass Task Force (2005) suggest that this figure could be significantly improved upon and that biomass could contribute more to the energy demands of the UK. It was concluded that in the future the UK could use biomass to satisfy about 4.5% of its present total energy needs, and that this energy was best extracted by combustion, possibly by co-firing with coal in power stations. Much of this biomass is part of a cascade of usage, re-use, and recycling of biomass materials, including residues generated on farms and from forestry within the global framework. However crops specifically grown for energy, such as cereals (Bullion, 2003), oil seed rape (Martini and Schell, 1998), sugar cane (Goldemberg *et al.*, 2004) (see Fig. 9.4), and soybeans (Wu *et al.*, 2004) are considered to have the potential to supply the expanding biofuels market.

The major biomass-derived fuel is ethanol produced from the fermentation of sugars or starches. Three countries, Brazil, US, and India, produce 90% of the world's ethanol from biomass with the total ethanol production being equivalent to 0.02 TW. Brazil has invested significantly in producing ethanol from sugar cane which is used as a substitute or as an additive to gasoline (Geller, 1985). Although the production and use of this bio-ethanol has been heavily subsidized, improved technologies and the rising cost of petroleum means that ethanol fuels are now competing favourably with gasoline in Brazil, and are having an impact in some parts of the US and Canada as well as India and the Far East.

For thousands of years, biomass was the only primary energy source available to mankind. For the last two centuries, however, energy demand has outpaced biomass production. Although biomass can still contribute to this demand in many ways and to different extents depending on climate and available landmass, it is hard to see how it could match the present level of global fossil fuel consumption or to cope with the increasing demands for energy in the future. Even with the best known energy crops, a power target of 20 TW power would require covering about 30% of the land mass of Earth, which corresponds to almost three times all cultivatable land currently used for agriculture. To reduce this to a reasonable level so as not to seriously compete with global food production would require the biomass crops to have solar energy conversion efficiencies close to the theoretical maximum of about 4.5%. Nevertheless, biomass could make a significant contribution to global energy demand if special plant breeding or genetic engineering yielded a new generation of environmentally robust energy crops, requiring minimal inputs of water and fertilizer and able to convert solar energy at efficiencies well above 1%. New and improved technologies to extract this energy will also be important.

Finally we should remind ourselves that biomass is not just a store of energy, but is also a source of complex molecules that constitute our food and provide us with valuable materials such as timber, linen, cotton, oils, rubbers, sugars, and starches. The increased use of 'designer' plants to produce high value chemicals for chemical and pharmaceutical industries should not be underestimated. Moreover new methodologies will emerge to release the valuable molecular building blocks of cellulose, lignin, and other polymers which constitute plant cell walls and fibres. In these various ways plant biomass will contribute to the energy requirements of modern industrialized society.

Although it may be possible to engineer plants and organisms to become efficient energy converting 'machines' and 'chemical factories', the overall efficiency will rarely exceed 1% and will usually be much less. Moreover, the growing of energy crops on a very large scale will compete with the traditional use of cultivatable land for food production. However, there is an alternative and complementary approach for utilizing solar energy. It may be possible to develop highly efficient, all-artificial, molecular-level energy-converting technologies which exploit the principles of natural photosynthesis.

The photosynthetic water splitting apparatus

As explained earlier, the process of oxygenic photosynthesis is underpinned by the light driven water splitting reaction that occurs in an enzyme found in plants, algae, and cyanobacteria known as PSII (see Fig. 9.4 below). Solar energy is absorbed by chlorophyll and other pigments and is transferred efficiently to the

PSII reaction centre where a process called charge separation takes place. This initial conversion of light energy into electrical energy occurs at the maximum thermodynamic efficiency of greater than 70%, and generates a radical pair state $P680^+ Pheo^-$ where P680 is a chlorophyll *a* molecule and Pheo is a pheophytin *a* molecule (chlorophyll molecule without a Mg ion ligated into its tetrapyrrole head group). The essential point is that the energy of the photon has been used to remove an electron from one site (P680) and place it on another (Pheo). The redox potential of the oxidized intermediate $P680^+$ is estimated to be more than $+1$ V (so it is a powerful oxidant), while that of $Pheo^-$ is about -0.5 V (so it is a powerful reducing agent that could in principle convert water to hydrogen). The electron could transfer back to P680, rendering the initial photoexcitation futile, but this does not normally occur. Neither is H_2 produced; rather, each electron is transported along a special 'relay' of electron carriers, involving iron, copper, and organic molecules such as plastoquinol (PQH_2) to a second photosystem PSI (see Fig. 9.2), where it is received by another oxidized chlorophyll molecule, known as $P700^{\cdot+}$. The result is to generate an even more powerful reducing agent (the redox potential is approximately -0.7 V or more negative still). In this way sufficient energy is accumulated to drive the fixation of carbon dioxide: the production of sugars not only requires the generation of the reduced 'hydrogen carrier', nicotinamide adenine dinucleotide phosphate (NADPH), but also requires the energy rich molecule adenosine triphosphate (ATP) formed by the release of some energy during electron transfer from PSII to PSI (in the form of an electrochemical potential gradient of protons).

We have yet to address the fate of the highly oxidizing intermediate $P680^{\cdot+}$ which must receive an electron in order to return to its resting state and be ready for the next photoexcitation. The beautiful feature now is that these electrons are drawn from water, resulting in its oxidation to O_2, hence the term *oxygenic* photosynthesis. Clean conversion of two H_2O molecules to O_2 is a difficult reaction that is achieved by a special catalytic centre—a cluster of four manganese (Mn) ions and a calcium ion (Ca^{2+}).

Overall, the splitting of water into O_2 and the equivalent of $2H_2$ (two NADPH) requires four electrons and PSII must absorb four photons ($4h\nu$) to drive this reaction:

$$2H_2O \xrightarrow{4h\nu} O_2 + 4H^+ \text{ and } 4e^-$$

The light-induced electrical charge separation events occurring in PSII, PSI, and related reaction centres in anoxygenic photosynthetic bacteria (organisms that do not split water) are highly efficient. The organization of the electron carriers in these nanomolecular photovoltaic devices are optimized to facilitate forward energy storing reactions and minimizing backward, energy wasting reactions. There is considerable information about these photosynthetic reaction centres which indicates that they are structurally and functionally very similar. Indeed, there are aspects of their design which could be incorporated into an 'artificial photosynthetic' system and are similar to existing photovoltaic systems, particularly the photoelectrochemical solar cells described in Chapter 8.

Similarly, the light harvesting systems of different photosynthetic organisms have common principles for facilitating energy capture across the whole of the visible spectrum, and aiding efficient energy transfer to the associated reaction centres with minimum losses of energy. Again, detailed spectroscopic and structural studies have revealed the molecular basis of these systems, details which could also be adopted for designing light concentrating systems for a new generation of solar energy converting technologies.

However it is the water splitting reaction of PSII which holds the greatest promise for developing new technologies for converting solar radiation into usable energy, particularly in generating H_2. Oxygenic photosynthesis is believed to have evolved about 2.5 billion years ago and was the 'big bang of evolution', since for the first time living organisms had available an 'inexhaustible' supply of an electron donor (water) to convert carbon dioxide into organic molecules. Furthermore, the waste product O_2 was eventually to become the standard oxidant for higher life forms that evolved the capabilities to deal with its oxidizing power. From that time onwards, life on Earth would prosper and diversify on an enormous scale; biology had solved its energy problem.

Clearly, using solar energy to split water to produce hydrogen rather than organic molecules is also the perfect solution for mankind. In principle, the technology exists today to do this. Electricity can be generated by photovoltaic solar cells and used to electrolyse water. With a solar cell efficiency of 10%, and 65% efficiency for the electrolysis system, the overall efficiency would be 6.5%. Electrolysis relies on electrodes that contain platinum or other catalysts which are in limited supply. At present very little hydrogen is generated by conventional electrolysis because of the high price of electricity generated by conventional means. Similarly, the cost of photovoltaic cells marginalizes this route for using solar energy to produce hydrogen. But perhaps a bio-inspired water splitting catalyst can be devised which works along similar chemical principles used by PSII.

Hydrogen cannot be obtained from water without simultaneous production of oxygen, which, because it involves four electrons rather than two, is a more complex reaction and occurs with greater difficulty. Because of the importance of understanding the chemistry of the water splitting reaction of PSII, there has been a battery of techniques employed to probe the molecular mechanisms involved and to investigate the structure of the catalytic centre.

Recently the crystal structure of PSII has been obtained by X-ray diffraction analysis, revealing the organization of the atoms in the Mn—cluster and details

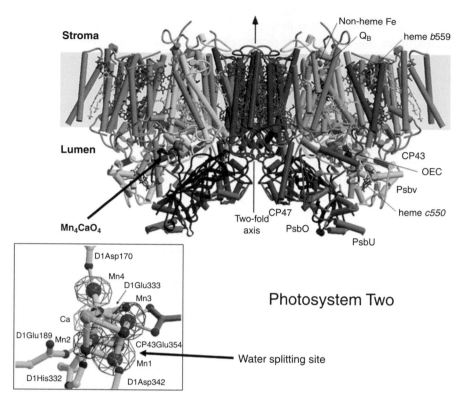

Photosystem Two

Figure 9.4 Structure of Photosystem II, the water splitting enzyme of photosynthesis, determined by X-ray crystallography (Ferreira *et al.*, 2004, Barber *et al.*, 2004). The complex spans the thylakoid membrane of cyanobacteria, plant, and algal chloroplasts. It comprises two monomers related to each other by a two-fold axis. Each monomer contains 19 different protein subunits (16 of which are located within the membrane spanning region and consist mainly of α-helices depicted by cylinders), and 57 cofactors including 36 chlorophyll a molecules. The water splitting site (inset) consists of a $Mn_3Ca^{2C}O_4$ cluster with a fourth Mn linked to the cubane by a bridging oxide ion. This catalytic centre is buried on the lumenal side of the complex.

of its protein environment (Ferreira *et al.*, 2004; Barber *et al.*, 2004). The PSII complex is isolated from a cyanobacterium called *Thermosynechococcus elongatus*. It contains 19 different protein subunits with the reaction centre, composed of the D1 and D2 proteins, at its heart (see Fig. 9.4).

The crystal structure of PSII revealed the organization of the cofactors involved in primary charge separation in the reaction centre (Fig. 9.5). However the most important outcome of this structural study was the conclusion that the water splitting centre consists of a cubane-like structure containing three manganese and one calcium linked together by oxide ions. The fourth manganese is linked to the cubane by a single oxide bridge (see Fig. 9.4, insert). Surrounding the

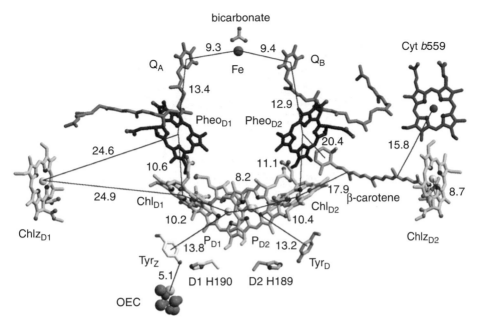

bicarbonate

Figure 9.5   Organization of the electron transfer cofactors in the reaction centre of Photosystem II as revealed by X-ray crystallography (Ferreira *et al.*, 2004; Barber *et al.*, 2004). Excitation of the reaction centre via the chlorophylls (Chl) leads to electron transfer from the special chlorophylls called P_{D1} and P_{D2} to the pheophytin (Pheo) acceptor leading to the radical pair $P680^{\cdot C} Pheo^{\cdot -}$. The radical cation of P680 is localized on P_{D1} while the radical anion is located on $Pheo_{D1}$. The electron on $Pheo_{D1}^{-}$ transfer rapidly to a firmly bound plastoquinone Q_A and then to a second plastoquinone Q_B. When the Q_B plastoquinol (PQH_2) diffuses from the Q_B-binding site into the lipid matrix of the membrane, $P680^{\cdot C}$ is reduced by a redox active tyrosine (Tyr_Z) which then extracts electrons from the $Mn_3Ca^{2C}O_4$ cluster (the oxygen-evolving centre (OEC)). These electron transfer processes occur mainly on the D1-side of the reaction centre and the symmetrically related cofactors located on the D2-side are mainly non-functional. Other cofactors, including the haem of cytochrome b_{559} (Cyt b_{559}) and the β-carotene molecule, help protect PSII against photoinduced damage.

$Mn_4Ca^{2+}O_4$ cluster are several amino acid residues that either provide ligands to the metal ions or act to facilitate hydrogen bonding networks, which almost certainly play a key role in the deprotonation of the H_2O molecules undergoing oxidation. A nearby tyrosine (residue 161 of the D1 protein) functions as an intermediate electron carrier between the $Mn_4Ca^{2+}O_4$−cluster and $P680^{\cdot +}$. Most of the key amino acids identified in the water splitting site belong to the D1 protein although another PSII protein, known as CP_{43}, also provides key residues. All these amino acids are fully conserved in all known amino acid sequences of the D1 protein and CP43 whether they are from prokaryotic cyanobacteria or eukaryotic algae and higher plants. It therefore appears highly likely that this catalytic centre is completely unique and unchanged throughout biology.

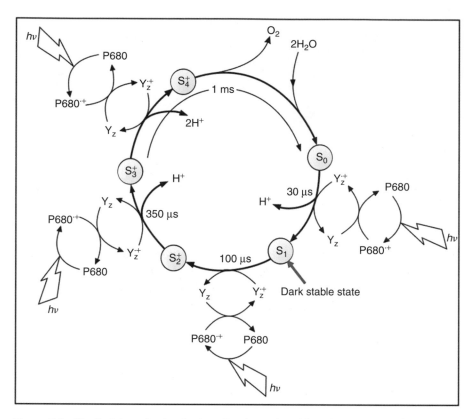

Figure 9.6 The S-state cycle showing how the absorption of four photons of light ($h\nu$) by P680 drives the splitting of two water molecules and formation of O_2 through a consecutive series of five intermediates (S_0, S_1, S_2, S_3, and S_4). Protons (H^C) are released during this cycle except for the S_1 to S_2 transition. Electron donation from the $Mn_3Ca^{2C}O_4$ cluster to $P680^{\cdot C}$ is aided by the redox active tyrosine Tyr_Z, abbreviated to Y_Z here. Also shown are half-times for the various steps of the cycle.

With this information, realistic chemical schemes are being proposed for the water splitting reaction. It has been known for some time that there are at least five intermediate states leading to the formation of O_2, known as S-states. The sequential advancement from S_0 to S_4 is driven by each photochemical turnover of the PSII reaction centre as depicted in the S-state cycle (Fig. 9.6). Progression through the S-states to S_4 builds up four oxidizing equivalents, which are reduced in the final step (S_4 to S_0) by four electrons derived from two substrate water molecules with the concomitant formation of dioxygen.

An important feature of manganese is its ability to exist in several different oxidation states, most of which are highly oxidizing. It is possible that in the S_4-state one of the Mn atoms (possibly the one that is not part of the cubane) has

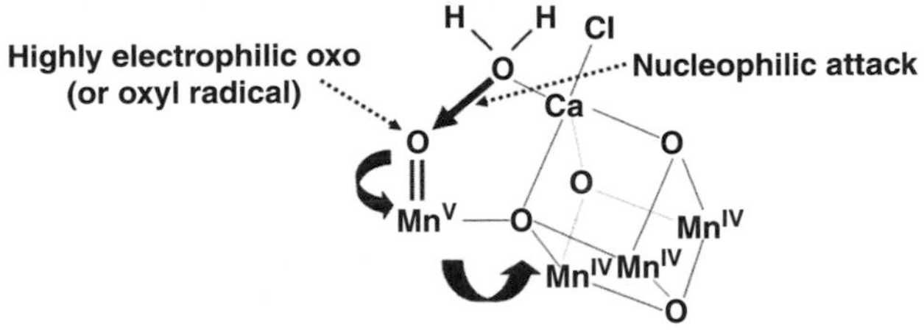

Figure 9.7   Formation of O_2 by photosynthesis—a plausible mechanism for formation of the O-O bond. The electron deficiency of the highly oxidized Mn-cluster causes deprotonation of a substrate water molecule that is bound at the remote Mn. This O-atom is then attacked by the more electron-rich oxygen atom of a second water molecule that is bound more loosely to Ca^{2C}. The arrows indicate direction of movement of electrons.

been oxidized to Mn(V) and the substrate water bound to it has been deprotonated to produce an oxide ion. This Mn(V) $=$ O, or possibly Mn(IV)-oxyl radical species (depending on how the electron distribution is described), is already very oxidizing and is linked to the three other Mn ion poised in high oxidation states (probably all Mn(IV)). The bound oxide will therefore be highly electron deficient and very electrophilic. This situation prompts attack by the oxygen atom of the second water molecule, bound to Ca^{2+}, which results in formation of an O-O bond and ultimately molecular oxygen (see Fig. 9.7). More details of how this reaction may occur have been formulated McEvoy and Brudvig (2004) and McEvoy *et al.*, (2005).

Artificial photosynthesis: a new technology

Although progress has been made in mimicking photosynthesis in artificial systems, researchers have not yet developed components that are both efficient and robust for incorporation into a working system for solar fuel production. To date, research has focused on design and synthesis of molecular systems able to mimic one part of this reaction—the light driven charge separation that occurs in photosynthetic reaction centres. The bio-inspired systems employ chromophores to absorb light energy, analogous with the photosynthetic pigments, such as chlorophyll. Often, however, the chromophores are directly engaged in the electron transfer processes and in this way act like the redox active chlorophylls within the photosynthetic reaction centres, for example P680 and P700. Having an antenna, or light harvesting array (which does not carry out charge separation itself) is an engineering design adopted by natural photosynthesis to maximize solar energy absorption. Arrays of light harvesting chromophores, able to funnel energy to a central site where charge separation occurs, have been demonstrated (Liddell *et al.*, 2004). However, these arrays are difficult to make and it would be desirable to create self-assembled functional antenna arrays using robust dyes such as those used as pigments in industrial paints.

The insights gleaned from the recent structural determination of PSII have major implications for the design of an artificial catalytic system for using solar energy to release hydrogen from water. The hydrogen produced could be used directly as an energy source but could also be used, as it is in photosynthesis, to reduce carbon dioxide to other types of fuels such as methane. Artificial catalysts reproducing the reactions of PSII may have to incorporate other bio-inspired features, particularly light harvesting systems to optimize energy capture over the whole visible spectrum. Our understanding of photosynthetic light harvesting is already at a sufficiently advanced level that chemical mimics can be designed.

The challenge is to have a molecular arrangement in the artificial catalyst that will allow the O-O bond to form. Recently it has been demonstrated that catalysts based on Mn are capable of water splitting, and generation of oxygen occurs when a strong oxidant is used to drive the Mn into high oxidation states (believed to be Mn(V)) (Limburg *et al.*, 1999). Light driven water splitting can, however, be accomplished using semiconductor-based photocatalysis. This was first demonstrated by Fujishima and Honda, who initially used TiO_2, although it was necessary to replace TiO_2 by $SrTiO_3$ in order to produce both hydrogen and oxygen. Since these reactions are driven by high energy UV radiation they are of limited practical use. Current efforts are being made to dope semiconductors with dyes able to carry out photo-driven redox reactions using visible light, as discussed in detail in Chapter 8. Similar types of semiconductor technologies are being developed to catalyse the photoelectrocatalytic reduction of carbon dioxide.

Policies and implementation

The enormous untapped potential of solar energy is an opportunity which should be addressed with urgency. Photosynthesis is the most successful solar energy converter on Earth. It provides energy for all life on our planet and is the source of the fossil fuels that drive our technologies. There is no reason why the chemical reactions devised by photosynthetic organisms cannot be mimicked by the ingenuity of humans. We already have a considerable knowledge to start from and the emerging nanotechnologies to exploit it further. With a concerted input of the talents of scientists trained in different disciplines it should be possible to move the technologies of solar energy cells forward.

The time has thus come to exploit our considerable knowledge of the molecular processes of photosynthesis and plant molecular biology to attack the challenge of providing non-polluting renewable energy for the future benefit of mankind. There are two avenues that should be explored with vigour:

1. *Biomass*—To continue to improve our knowledge of the molecular genetics which underlie plant metabolism, growth, and survival with a view to engineering designer crops with high solar energy conversion efficiencies but having minimum requirements for fertilizer, insecticides, and water. Genetic manipulation of the chemical composition of these robust energy plants will facilitate their use in a variety of ways, ranging from the production of liquid biofuels and biogas to high value chemicals for the chemical and pharmaceutical industries. Coupled with these objectives, there must be improvements of appropriate technologies to exploit biomass in a range of different ways. These developments will in part be driven by market forces, and in part by the policies of world governments and organizations towards reducing CO_2 emissions with the purpose of minimizing man-made global climate change.

2. *Artificial photosynthesis*—To establish a multidisciplinary effort to construct robust artificial systems able to exploit solar energy to split water with the capacity to produce fuels like methane. In recent years there has been important progress in understanding the molecular processes of photosynthetic energy conversion and water splitting. This knowledge base can now be combined with that of photovoltaics and nanotechnology to construct an 'artificial water splitting' system with a solar energy conversion efficiency of at least 10%. If artificial systems or related photovoltaic systems could convert solar energy at 10% efficiency, we would need to cover only 0.16% of the Earth's surface to satisfy global energy requirements of 20 TW (Lewis, 2005). Unlike a biological leaf this artificial equivalent could, for example, be placed in arid desert areas whose total areas well exceed that required (Fig. 9.8).

Because of this, the 'artificial leaf' will not compete for cultivatable land in the way that massive biomass production will. Since this is a long term goal with benefit to all, it will need investment by the international community in a way similar to that received for nuclear fusion research (see Chapter 7),

Finally, it is interesting to consider a quote from Jules Verne's novel *L'Ile Mysterieuse,* written in 1875, and compare this with the solution depicted in Fig. 9.9.

'I believe that water will one day be used as a fuel, because the hydrogen and oxygen which constitute it, used separately or together, will furnish an inexhaustible source of heat and light.

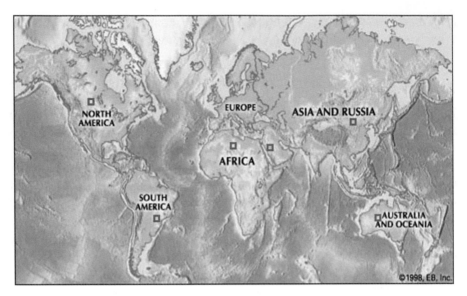

Figure 9.8   The six boxes represent the areas of land needed to obtain 20 TW of energy from solar radiation at 10% conversion efficiency. These areas together represent $5 \times 10^{11} m^2$ or 0.16% of the Earth's surface. Each box is equivalent to 3.3 TW. In practice, the sites would be smaller and distributed more evenly around the globe using, whenever possible, land that is not cultivatable (Source: Lewis, http://www.cce.caltech.edu:16080/faculty/lewis/research.html).

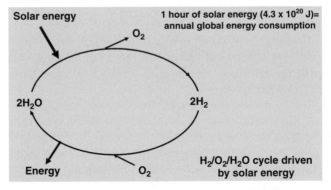

Figure 9.9   The ideal solution to the energy/CO_2 problem is to use solar energy to split water into O_2 and H_2. The H_2 can be used directly or indirectly as a fuel. This is essentially how biology solved its energy problem through the process of oxygenic photosynthesis and the secret of its success has been revealed in considerable detail. The time has come to construct catalysts which mimic the biological processes of water splitting, hydrogen production, and CO_2 extraction from the air. A major initiative would bring together scientists from a wide range of disciplines with the challenge of building an 'artificial leaf' with energy conversion efficiencies of 10% or more.

I therefore believe that, when coal deposits are oxidised, we will heat ourselves by means of water. Water is the coal of the future.'

Acknowledgements

I wish to thank the Biotechnology and Biological Science Research Council (BBSRC) for the financial support that led to the determination of the structure of PSII. I have drawn on information from a wide range of sources but I would particularly like to acknowledge that my thinking has been significantly influenced by Professor Nathan Lewis of Caltech via his website http://www.cce.caltech.edu:16080/faculty/lewis/research.html and Professor Daniel Nocera at MIT via discussion, both of whom are strong advocates of using solar energy as an inexhaustible and non-polluting source of energy for the long term benefit of mankind.

Resources and further information

Archer, M. D. and Barber, J. 2004, *Molecular to Global Photosynthesis*, Imperial College Press, London.

Barber, J. 2003. Photosystem II: The engine of life. *Quart. Revs. Biophys.* **36**, 71–89.

Barber, J. *et al.*, 2004. Structure of the oxygen evolving center of photosystem II and its mechanistic implications. *Phys. Chem. Chem. Phys.* **6**, 4737–42.

Blankenship, R. E. 2002. *Molecular Mechanisms of Photosynthesis*. Blackwell Science, Oxford.

Bolton, J.R. (1977) Solar Power and Fuels. Pub. Academic Press, New York, USA.

Bolton, J. R. 1979. Solar energy conversion in photosynthesis: features relevant to artificial systems for the photochemical conversion of solar energy. In *The Chemical Conversion and Storage of Solar Energy* (eds. King, J. B., Hautala, R. R. and Kutal, C. R.) Pub Humana Press, Clifton, N.J.

Bridgewater, A. V. and Maniatis, K. 2004. The production of biofuels by the thermochemical processing of biomass. In *Molecular to Global Photosynthesis*. (Ed. Archer, M. D. and Barber, J.) pp 521–611. Pub Imperial College Press, London.

Bullard, M. D. 2004. Photoconversion and energy crops. In *Molecular to Global Photosynthesis*. (Ed. Archer, M. D. and Barber, J.) pp 453–519. Pub Imperial College Press, London...

Bullion, A. 2003 Ethanol trends. *Int. Sugar J.* **1263**, 173–74.

Carbon Trust: Biomass Sector Review. London, UK 2005. hppt://www.thecarbontrust.co.uk/about/publications/BiomassSector-Final.pdf

European Technology Platform 'Plants for the Future' www.plantTP.com

Falkowsky, P. G. and Raven, J. A. 1997. *Aquatic Photosynthesis*. Blackwell, Oxford.

Ferreira, K. N., Iverson, T. M., Maghlaoui, K., Barber, J. and Iwata, S. 2004. Architecture of the photosynthetic oxygen evolving centre. *Science*, **303**, 1831–38.

Fujishima, A. and Honda, K. 1972. Electrochemical photolysis of water at a semiconductor electrode. *Nature,* **238**, 37–8.

Geller, H. S. 1985. Ethanol fuel from sugar cane in Brazil. *Ann. Rev. Energy*, **10**, 135–64.

Goldemberg, J. *et al.*, 2004. Ethanol learning curve: the Brazilian experience. *Biomass and Bioenergy*, **26**, 301–4.

Gust, D., Moore, T. A. and Moore, A. L. 2001. Mimicking photosynthetic solar energy transduction. *Acc. Chem. Res*. **34**, 40–8.

Heyduk, A. F. and Nocera, D. G. 2001. Hydrogen produced from hydrohalic acid solutions by a two electron mixed valence photocatalyst. *Science,* **293**, 1639–41.

Hirose, T., Maeno, Y. and Himeda, Y. J. 2003. Photocatalytic carbon dioxide photoreduction by $Co(bpy)_3^{2+}$ sensitized by $Ru(bpy)_3^{2+}$ fixed to cation exchange polymer. *J. Mol. Catal. A: Chem.* **193**, 27–32.

Hoffert, M. T. *et al.* 1998. Energy implications of future stabilization of atmospheric CO_2. *Nature*, **395**, 881–84.

Hoffert, M. T. *et al.* 2002. Advanced technology paths to global climate stability: energy for a greenhouse planet. *Science.* **298**, 981–87.

Houghton, J. T., Ding, Y., Griggs, D. J., Noguer, M., van der Linden, P. J. and Xiaosu, D.(eds) 2001. *Climate Change: The Scientific Basis,* p. 944 Pub Cambridge University Press, Cambridge, UK.

Imahri, H. J., Mori, Y., and Matano, Y. Nanostructured artificial photosynthesis. *J. Photochem. Photobiol. C: Photochem Rev*. **4**, 51–83.

Kato, H., Asakura, K. and Kudo, A. 2003. Highly efficient water splitting into H_2 and O_2 over lanthanum-doped $NaTaO_3$ photocatalysts with high crystallinity and surface nanostructure. *J. Am. Chem. Soc.* **125**, 3082–89.

Lewis, N. S. 2005. *Chemical Challenge in Renewable Energy.* Lecture text on http://www.cce.caltech.edu:16080/faculty/lewis/research.html

Liddell, P. A. *et al.* 2004. Photoinduced electron transfer in a symmetrical diporyphyrin-fullerene triad. *Phys. Chem. Chem. Phys.* **6**, 5509–5515.

Limberg, J., Vrettos, J. S., Liable-Sands, L. M., Rheingold, A. L., Crabtree, R. H. and Brudvig, G. W. 1999. A functional model for O-O bond formation by the oxygen evolving complex in photosystem II. *Science,* **283**, 1524–27.

Martini, N. and Schell, J. S. 1998. *Plant Oils as Fuel: Present State of Science and Future Developments.* Springer-Verlag, Berlin.

McEvoy, J. P., Gascon, J. A., Batista, V. S. and Brudvig, G. W. 2005. The mechanism of photosynthetic water splitting. *Photochem. Photobiol. Sci.* **4**, 940–49.

McEvoy, J. P. and Brudvig, G. W. 2004. Structure-based mechanism of photosynthetic water oxidation. *Phys. Chem. Chem. Phys.* **6**, 4754–63.

Oogwijk, M., van dem Broek R., Berndes, G. And Faaij, A. 2001 A review of assessments on the future contribution of biomass energy. In *Proc. 1ˢᵗ World Conf. on Biomass for Energy and Industry.* Vol 1 pp. 296–99 James and James, London.

Parika, M. 2004 Global biomass fuel resources. *Biomass and Bioenergy*, **27**, 613–20.

Slesser, M. and Lewis, C. 1979 *Biological Energy Resources,* E & F.N. Spon, London.

Somerville, C. *et al.* 2004 Towards a systems approach to understand plant cell walls. *Science,* **306**, 2206–211.

Spath, P. L. and Drayton, D. C. 2003. Preliminary screening-technical and economic assessment of synthesis gas to fuels and chemicals and emphasis on the potential for biomass-derived syngas. National Renewable Energy Laboratory, Golden, Colo., NREL/TP-510-34929 p.160.

Thorndike, E.H. 1996. *Energy and the Environment*, Addison-Wesley, Reading, Mass.,

US Department of Energy (DOE) 2006. *Multi-year Technical Plan*, Washington D.C., USA http://devafdc:nrel.gov/biogeneral/Program_Review/MYTP.pdf

US Department of Energy (DOE). 2005. *Basic Research Needs for Solar Energy Utilization*. Report of the basic energy sciences workshop on solar energy utilization. http://www.sc.doe.gov/bes/reports/files/SEU_rpt.pdf

Walker, D. A. 1979. *Energy, Plants and Man*. Pub Packard Publishing Ltd., Chichester, UK. Carbon Trust (2005) Biomass Sector Review Pub Carbon Trust

World Energy Council 2001. *Survey of Energy Resources*. World Energy Council (www.worldenergy.org)

The author

Professor James Barber graduated in Chemistry from the University of Wales. He joined the Academic Staff of Imperial College in 1968 after completing a post-doctoral fellowship as the Unilever European Fellow at the State University of Leiden in The Netherlands. He was awarded an honorary doctorate of Stockholm University and elected a member of the European Academy 'Academia Europaea (1989). He has been Dean of the Royal College of Science at Imperial College and was Head of the Biochemistry Department for ten years (1989–1999). He occupies the Chair named after the Nobel Laureate Ernst Chain (co-discoverer of Penicillin). He has published over 400 original research papers and reviews in the field of plant biochemistry, has edited 16 specialized books, and was awarded the prestigious Flintoff Medal by the Royal Society of Chemistry in 2002. He was elected a foreign member of the Royal Swedish Academy of Sciences in 2003 and elected a Fellow of the Royal Society in 2005, and has recently been awarded the 2006 Biochemical Society Novartis medal and prize and the 2007 Wheland Medal and Prize from the University of Chicago. The core of his research has been to investigate photosynthesis and the functional role of the photosystems with emphasis on their structures. Much of his work has focused on Photosystem Two, a remarkable biological machine able to use light energy to split water into oxygen and reducing equivalents.

10. *Sustainable hydrogen energy*

Peter P. Edwards, Vladimir L. Kuznetsov,
and William I. F. David

Introduction

Hydrogen is the simplest and most abundant chemical element in our universe—it is the power source that fuels the Sun and its oxide forms the oceans that cover three quarters of our planet. This ubiquitous element could be part of our urgent quest for a cleaner, greener future. Hydrogen, in association with fuel cells, is widely considered to be pivotal to our world's energy requirements for the twenty-first century and it could potentially redefine the future global energy economy by replacing a carbon-based fossil fuel energy economy. The principal drivers behind the sustainable hydrogen energy vision are therefore:

- the urgent need for a reduction in global carbon dioxide emissions;

- the improvement of urban (local) air quality;

- the abiding concerns about the long-term viability of fossil fuel resources and the security of our energy supply;

- the creation of a new industrial and technological energy base—a base for innovation in the science and technology of a hydrogen/fuel cell energy landscape.

The ultimate realization of a hydrogen-based economy could confer enormous environmental and economic benefits, together with enhanced security of energy supply. However, the transition from a carbon-based (fossil fuel) energy system to a hydrogen-based economy involves significant scientific, technological, and socio-economic barriers. These include:

- low-carbon hydrogen production from clean or renewable sources;

- low-cost hydrogen storage;

- low-cost fuel cells;

- large-scale supporting infrastructure, and

- perceived safety problems.

In the present chapter we outline the basis of the growing worldwide interest in hydrogen energy and examine some of the important issues relating to the future development of hydrogen as an energy vector. As a 'snapshot' of international activity, we note, for example, that Japan regards the development and dissemination of fuel cells and hydrogen technologies as essential: the Ministry of Economy and Industry (METI) has set numerical targets of 5 million fuel cell vehicles and 10 million kW for the total power generation by stationary fuel cells by 2020. To meet these targets, METI has allocated an annual budget of some £150 million over four years.

Our assumption, therefore, is that ultimately a significant proportion of the world's future energy needs will be met by hydrogen, and this hydrogen will be used to power fuel cells.

Hydrogen and electricity: energy carriers

Unlike coal, gas, or oil, hydrogen is not a primary energy source which can be mined at source. Rather, its role more closely mirrors that of electricity as a secondary 'energy carrier' which must first be produced, using energy from another source, and then transported for future use where its latent chemical energy can be fully realized. Hydrogen, however, has a major advantage over electricity in that it can be stored as a chemical fuel and converted into energy using fuel cells or internal combustion engines and turbines. The only by-product is water at the point of use.

Importantly, hydrogen can also be used as a storage medium for electricity generated from intermittent, renewable resources such as solar, wind, wave and tidal power, and biomass; in this case hydrogen is not produced, but rather harvested from nature! This attractive 'energy-carrier' facet of hydrogen provides a realizable solution to one of the major issues of sustainable energy, namely, the vexed problem of intermittency of supply—the Sun does not always shine and the wind does not always blow! Hydrogen can naturally fulfil the vital storage function of smoothing the daily and seasonal fluctuations of renewable energy resources. It is this key element of the intrinsic energy storage capacity of hydrogen which provides the potent link between sustainable energy technologies and any sustainable energy economy, generally placed under the umbrella of a 'hydrogen economy'.

It is recognized that hydrogen as an energy carrier or vector has the most potential to effect a radical change to our energy system. The implications for both the environment and energy security depend on the source of hydrogen—the benefits are obviously most significant if the hydrogen is manufactured from these sustainable sources noted above.

The importance of hydrogen as a potential energy carrier has also increased significantly over the last decade because of rapid advances in fuel cell technology. A main avenue of activity has been in the transportation sector where most of the world's major vehicle manufacturers are investing heavily in fuel cell vehicle R&D programmes. A hydrogen fuel cell is a device akin to a continuously recharging battery; a fuel cell generates electricity by the electrochemical reaction of hydrogen and oxygen from the air (Chapter 11). While batteries store energy, fuel cells can produce electricity *continuously* as long as fuel and air are supplied.

Hydrogen fuel cells consist of two electrodes (anode and cathode) separated, for example, by a polymer electrolyte membrane (Fig. 10.1). Hydrogen or a hydrogen-containing fuel (e.g. hydrocarbons) and oxygen are fed into the anode and cathode of the fuel cell and the electrochemical reactions to yield water and electrical energy are assisted by catalysts take place at the electrodes. The role of the electrolyte is to enables the transport of ions between the electrodes while the excess electrons flow through an external circuit to provide an electrical current.

Any hydrogen-rich fuel can be used in various types of fuel cells but using a hydrocarbon-based fuel inevitably leads to a carbon dioxide emission. Hydrogen-powered fuel cells emit only water and have virtually no pollutant emissions, not even nitrogen oxides, since they operate at temperatures that are much lower than internal combustion engines. A fuel cell can convert hydrogen into electricity two or three times more efficiently than internal combustion engines or turbines and produce much less noise (a feature utilized in their operations in submarines). In transport, hydrogen fuel cell engines operate at an efficiency of up to 65%, compared to 25% for present-day petrol driven car engines. When heat generated in fuel cells is also utilized in Combined Heat and Power (CHP) systems, an overall efficiency in excess of 85% can be achieved. Fuel cells are also remarkably

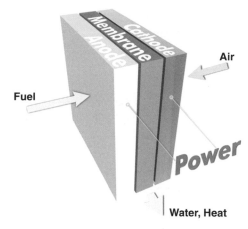

Figure 10.1 A schematic representation of a fuel cell. Air and fuel combine to generate power and water (as the exhaust gas) at the point of use. Courtesy of Karl Harrison, University of Oxford.

versatile and their functions are 'scalable' so that they can be used to provide electricity in applications ranging from a laptop computer to large-scale CHP units for major urban uses.

In brief, the synergetic complementarity of hydrogen and electricity represent one of the most promising routes to a sustainable energy future, and fuel cells provide, arguably, the most efficient conversion device for converting hydrogen and other fuels into electricity.

Hydrogen-fuelled fuel cell vehicles are increasingly seen as an attractive alternative to other zero-emission vehicles such as battery driven electric cars, because the chemical energy density of hydrogen is significantly higher than that found in electric battery materials (Winter and Brodd, 2004). Hydrogen fuel cells could deliver a much longer operational life-time than that of electric batteries, and the same high specific energy as traditional combustion engines. However, various major technological hurdles must be overcome before fuel cells can compete effectively, in terms of overall performance and cost, with fossil-fuel based internal combustion engines in automotive applications. Fuel cell cars, currently the focus of intense development activity worldwide, are not expected to reach mass market until 2015, or indeed beyond.

Fig. 10.2 illustrates the central role of hydrogen as an energy carrier linking multiple hydrogen production methods and hydrogen storage and various end user applications. One of the principal advantages of hydrogen as an energy carrier is obviously the diversity of production methods from a variety of domestic resources (Fig. 10.2). A typical energy chain for hydrogen comprises hydrogen production, distribution, and delivery through hydrogen storage and ultimately

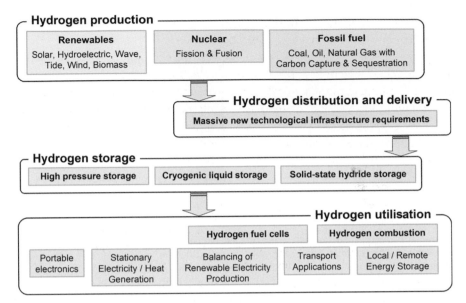

Figure 10.2   Hydrogen as an energy carrier linking hydrogen production methods to the end users.

its utilization. The energy chain for sustainable hydrogen energy would comprise the harvesting of sunlight or other energy sources to yield hydrogen as the energy carrier, the storage and distribution of this energy carrier to its utilisation at an end device—centred on either fuel cells or combustion—where it is converted to power.

At the present time, there are three major technological barriers that must be overcome for any transition from a carbon-based (fossil fuel) energy system to a hydrogen-based economy. First, the cost of efficient and sustainable hydrogen production and delivery for the large amount of *extra* hydrogen for fuel cell activities must be significantly reduced. Second, new generation of hydrogen storage systems for both vehicular and stationary applications must be developed. Finally, the cost of fuel cells systems must be dramatically reduced and their durability must be improved. In addition, pivotal issues relating to the socio-economic aspects of any transition to a sustainable hydrogen economy must be examined as parallel, indeed coupled, activities (McDowall and Eames, 2005).

Hydrogen production

Even though hydrogen is the third most abundant chemical element in the Earth's crust, almost all of this hydrogen is bound up in chemical compounds with other elements. It must, therefore, be produced from other hydrogen-containing sources using the input of primary energy such as electricity or heat.

Hydrogen can be produced from coal, natural gas, and other hydrocarbons by a variety of techniques, from water by electrolysis, photolytic splitting, or high temperature thermochemical cycles—from biomass and municipal waste by fermentation, gasification, or pyrolysis. Such a diversity of production sources contributes significantly to security of fuel supply.

At present, the vast majority of the world's hydrogen (so-called 'merchant hydrogen') is ironically produced from fossil fuels by steam reforming of natural gas and partial oxidation of coal or heavy hydrocarbons. These methods can take advantage of economies of scale and are currently the cheapest and most established techniques of producing hydrogen. They can be used in the short to middle term to meet hydrogen fuel demand and enable the proving and testing of technologies relating to hydrogen storage, distribution, safety, and use. However, in the long term it is clearly unsustainable that the hydrogen economy is driven by the carbon economy. At this point, it is important to appreciate that most of the world's hydrogen is currently produced for ammonia/fertilizer synthesis and not easily committed to any new major market (e.g. for hydrogen/fuel cell activities). In addition, the production of hydrogen from fossil fuels using reforming and gasification processes always leads to the emission of CO_2, the principal cause of global climate change. Carbon dioxide emissions can be efficiently managed at large-scale facilities through so-called carbon dioxide sequestration, which involves the capture, liquefaction, transport, and injection of liquid carbon dioxide underground (e.g. into depleted natural gas and oil wells or geological formations). All the operations associated with sequestration are energy intensive, costly, and potentially damaging to our environment. The key risk results from the uncertain long-term ecological consequences of carbon dioxide sequestration.

To achieve the benefits of a truly sustainable hydrogen energy economy, we must clearly move to a situation where hydrogen is produced by electrolysis or splitting of non-fossil resources, such as water, using (ideally) electricity derived from renewable energy sources. The holy grail of hydrogen production is therefore the efficient, direct conversion of sunlight through a photochemical process that utilizes solar energy to split water directly to its constituents, hydrogen and oxygen, without the use of electricity. This requires innovative materials discovery and solar cell development. The ideal production route then harvests 'solar hydrogen', the power of the Sun, to split water from our oceans. Solar photodecomposition of water has been cited as the only major—but long-term—solution to a CO_2-free route for the mass production of the huge volumes of H_2 needed if the hydrogen economy is to emerge.

Current nuclear (fission) technology generates electricity that can be used to produce hydrogen by the electrolysis of water. Advanced nuclear reactors are also being developed that will enable high-temperature water electrolysis (with less electrical energy needed) or thermochemical cycles that will use heat and a chemical process to dissociate water. Fusion power, if successfully developed, could be a clean, abundant, and carbon-free resource for hydrogen production.

Until 2020, hydrogen production from fossil fuels and by electrolysis of water using grid electricity is expected to be the most important sources of hydrogen. During this transition period, advanced and clean reformation/gasification processes, carbon dioxide capture and sequestration, and new efficient and low cost electrolysers will have to be developed. However, in the long term, sustainable hydrogen production technologies based on renewable energy resources should become commercially competitive and gradually replace the fossil fuel reformation/gasification. Hydrogen, produced by solar photodecomposition of water or by electrolysis of water using electricity generated from renewable resources, has the potential to be the clean, sustainable and, therefore, climate-neutral energy carrier of the future, eventually eliminating greenhouse gas emissions from the energy sector.

The use of biological processes to produce hydrogen is clearly attractive if one could indeed demonstrate that any such approach could be used to produce the huge volumes of hydrogen. Of course, this is a potentially CO_2-free route if no fertilisers are utilized in the feedstock.

The use of hydrogen-fuelled vehicles will also depend on the successful development of a widespread refuelling infrastructure. The components of a national hydrogen delivery and distribution network (including hydrogen pipelines) will need to be developed to provide a reliable supply of low-cost hydrogen to consumers. In the past, the national infrastructure has been constructed by monopolies and public bodies. A major area of activity for the UK, and indeed all countries, will be to develop scenarios as to how this hydrogen infrastructure will arise.

Hydrogen storage

One of the crucial technological barriers to the widespread use of hydrogen is the lack of a safe, low-weight, and low-cost storage method for hydrogen with a high energy density (Harris *et al.*, 2004; Crabtree *et al.*, 2004). Hydrogen contains more energy on a weight-for-weight basis than any other substance. Unfortunately, since it is the lightest chemical element in the Periodic Table it also has a very low energy density per unit volume (Table 10.1).

Traditional storage options for hydrogen have centred upon high-pressure (up to 700 bar) gas containers or cryogenically cooled (liquefied) fluid hydrogen. One

Table 10.1 Gravimetric (specific energy) and volumetric (energy density) energy content of various fuels, hydrogen storage options and energy sources (container weight and volume are excluded).

Fuel	Specific energy (kWh/kg)	Energy density (kWh/dm^3)
Liquid hydrogen	33.3	2.37
Hydrogen (200 bar)	33.3	0.53
Liquid natural gas	13.9	5.6
Natural gas (200 bar)	13.9	2.3
Petrol	12.8	9.5
Diesel	12.6	10.6
Coal	8.2	7.6
LiBH$_4$	6.16	4.0
Methanol	5.5	4.4
Wood	4.2	3.0
Electricity (Li-ion battery)	0.55	1.7

downside to these methods is a significant energy penalty—up to 20% of the energy content of hydrogen is required to compress the gas and up to 40% to liquefy it. Nevertheless, incremental technology developments are gradually reducing these energy penalties. A recent US National Academy of Sciences Report (US Department of Energy, 2004), however, concludes that both liquid and compressed hydrogen storage show little promise of long-term practicality. Another crucial issue that confronts the use of high-pressure and cryogenic storage, centres on public perception and acceptability associated with the use of pressurized gas and liquid hydrogen containment. On-board storage of hydrogen is a formidable scientific and technological problem.

Hydrogen storage requires a major technological breakthrough and this is likely to occur in the most viable alternative to compressed and liquid hydrogen, namely the storage of hydrogen in solids or liquids. Metallic hydrides, for example, can safely and effectively store hydrogen within their crystal structure. Hydrogen is first 'sorbed' into the material and is released under controlled heating of the solid. The development of new solid-state hydrogen storage materials could herald a breakthrough in the technology of hydrogen storage and would have a major impact on the transition to a hydrogen economy (Harris *et al.*, 2004; Crabtree *et al.*, 2004) .

The challenge of finding a suitable hydrogen storage media is illustrated in Fig. 10.3 which displays the gravimetric and volumetric energy densities of hydrogen stored using various storage methods. It is seen that neither cryogenic

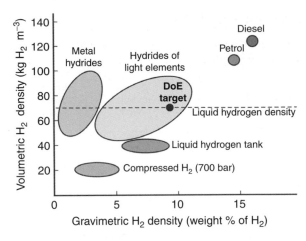

Figure 10.3   Gravimetric and volumetric densities of various hydrogen storage options (including weight and volume of the storage container). 'DoE target' represents the US Department of Energy target set for an 'ideal' hydrogen storage material. Metal hydrides are conventional, heavy metal hydrides such as LaNi$_5$ etc.

nor high-pressure hydrogen storage options can meet the mid-term targets for transport use. It is becoming increasingly accepted that solid state hydrogen storage using ionic-covalent hydrides of light elements, such as lithium, boron, sodium, magnesium, and aluminium (or some combination of these elements) represents the only method able to achieve the gravimetric and volumetric target densities.

For hydrogen-fuel cell transportation use—widely regarded as the first major inroad into the hydrogen economy—a suitable material for on-board storage should be able to store a high weight percent and high volume density of hydrogen and, equally important, rapidly discharge and charge this same amount of hydrogen at acceptable operating temperatures (typically around 50—100°C). This represents a particular challenging set of credentials for an ideal storage material; at present no known material meets these critical requirements. It is widely accepted that a solution to the hydrogen storage problem is the key to the transition to a hydrogen economy and large financial resources, especially in the US and Japan, are being channelled into this activity—'The Grand Challenge of Hydrogen Storage'. This key area represents the major thrust of our group's research activities at Oxford University and the Rutherford Appleton Laboratory. The resolution of the on-board storage problem has a potential 'game-changing' for transport and it is clear that progress can only be accelerated via multi-institution/multi-national coupled activities of the major international agencies of the IPHE (http://www.iphe.net/) and IEA (http://www.iea.org/).

A sustainable energy future

Our future energy choices are now inextricably tied to the fate of our global climate. The social and political will to improve our global climate, coupled with the need for a diverse, sustainable energy supply, are the major driving forces behind the hydrogen energy vision briefly outlined here. To achieve this vision within a realistic timeframe will require sustained scientific and technological innovation, together with a continued social and political commitment, and strong, coupled international activities and collaboration.

By 2050, the global energy demand could double or triple, and oil and gas supply is unlikely to be able to meet this demand. Hydrogen and fuel cells are considered in many countries as an important alternative energy vector and a key technology for future sustainable energy systems in the stationary power, transport, industrial, and residential sectors (European Commission, 2003; US Department of Energy, 2004). However, as with any major changes in the energy industry, the transition to a hydrogen economy will require several decades.

The timescale and evolution of such a transition is the focus of many 'Roadmaps' emanating from the USA, Japan, Canada, and the EU (amongst many others). For example, the European Commission has endorsed the concept of a Hydrogen and Fuel Cell Technology Platform with the expenditure of 2.8 billion € over a period of ten years. The introduction of hydrogen as an energy carrier has been identified as a possible strategy for moving the UK towards its voluntary adopted targets for CO_2 reduction of 60% of current levels by 2050 (DTI, 2003).

Table 10.2. summarizes the resulting forecasts of several roadmaps for deployment status, and targets for hydrogen technologies and fuel cell applications.

Table 10.2 Key assumptions on hydrogen and fuel cell applications.

Technology	Today	2020–2025	2050
Carbon capture and sequestration (CSS) (€/ton CO_2)	20–30	4–8	3–6
Hydrogen produced from coal with CCS (€/GJ)	8–10	7–9	3–5
Hydrogen transportation/storage cost (pipeline, 5,000 kg/h, 800 km) (€/GJ)	10–15	3	2
Hydrogen fuel cells (€/kW)	6,000–8,000	400	40
EU, Fuel cell vehicles, sold per year	n.a.	0.4–1.8 million	n.a.
Japan, Fuel cell vehicles, cumulative sale target	n.a.	5 million	n.a.
IEA forecast, global fleet of fuel cell vehicles	n.a.	n.a.	700 million

Source: International Energy Agency, 2006; European Hydrogen and Fuel Call Technology Platform, 2005.

At present, there is clearly a huge knowledge and technology gap separating us from the hydrogen economy. To achieve a significant penetration of hydrogen into future energy systems the methods of hydrogen production, distribution, storage, and utilization must be dramatically improved beyond their present performance, reliability, and cost.

Of course, hydrogen on its own cannot solve all of the complex energy problems facing our world today. However, it can provide a major alternative which attempts to shift our carbon-based global energy economy away from our dependence on rapidly depleting supplies of oil, ultimately towards a clean, renewable hydrogen energy future.

Conclusions

The development of hydrogen storage and fuel cell technologies is set to play a central role in addressing growing concerns over carbon emissions and climate change, as well as the future availability and security of energy supply. A recent study commissioned by the DTI found that hydrogen energy offers the prospect of meeting key UK policy goals for a sustainable energy future (E4tech, 2004). Together, hydrogen and fuel cells have the capability of producing a green revolution in transport by removing carbon dioxide emissions completely. Across the full range of energy use, these technologies provide a major opportunity to shift our carbon-based global energy economy ultimately to a clean, renewable and sustainable economy based on hydrogen.

The challenges are substantial and require scientific breakthroughs and significant technological developments, coupled with a continued social and political commitment. The widespread use of petroleum in the twentieth century was an attractive answer to a significant problem. But petroleum's success has created problems with pollution, global warming, energy security, and environmental impacts. In the twenty-first century, hydrogen now represents the attractive answer to these significant problems. Creating a new energy economy— and one that no longer centres on carbon fuels—will require the best thinking from the brightest minds. There is still a long road to travel before a true hydrogen energy revolution can occur—this will be a compelling and exciting journey!

Resources and further information

Committee on Alternatives and Strategies for Further Hydrogen Production and Use, National Research Council, National Academy of Engineering, 2004, *The Hydrogen Economy: Opportunities, Costs, Barriers, and R&D Needs,* The National Academies Press, Washington, D.C. Available from: http://www.nap.edu/books/0309091632/html/

Crabtree, G. W., Dresselhaus, M. S., and Buchanan, M.V. 2004. The hydrogen economy, *Physics Today*, **57**, 12, 39–44.

European Commission, 2003. *Hydrogen Energy and Fuel Cells A Vision of Our Future*, available from: http://www.europa.eu.int/comm/research/energy/pdf/hydrogen-report_en.pdf

E4tech, Element Energy, Eoin Lees Energy. December 2004. *A strategic framework for hydrogen energy in the UK*, Final report to the Department of Trade and Industry, UK, available from: http://www.dti.gov.uk/energy/sources/sustainable/hydrogen/strategic-framework/page26734.html

Harris, I. R., Book, D., Anderson, P. A. and Edwards, P. P. 2004 Hydrogen storage: the grand challenge, *The Fuel Cell Review*, June/July, 17–23.

International Energy Agency, 2006. *Hydrogen Production and Storage, R&D Priorities and Gaps*, available from: http://www.iea.org/Textbase/papers/2006/hydrogen.pdf

McDowall W. and Eames M., 2005. *Report of the September 2005 UK-SHEC Hydrogen Transitions Workshop*, PSI, London, available from: http://www.psi.org.uk/ukshec/pdf/11_Transition%20Workshop%20Report.pdf

The Advisory Council of the Hydrogen and Fuel Cell Technology Platform 2005, *European Hydrogen & Fuel Cell Technology Platform. Deployment Strategy*, available from: https://www.hfpeurope.org/hfp/keydocs

The Department of Trade and Industry,2003. *Energy White Paper*, UK, available from: http://www.dti.gov.uk/files/file10719.pdf

U.S. Department of Energy, 2004. *Hydrogen Posture Plan*. Washington, DC, available from: http://www.hydrogen.energy.gov/

Winter, M. and Brodd, R. J. 2004. What are batteries, fuel cells, and supercapacitors?, *Chemical Reviews*, **104**, 4245–69.

The authors

Peter P. Edwards F.R.S. is Professor and Head of Inorganic Chemistry at the University of Oxford. In 1996 he was elected Fellow of the Royal Society and in 2003 was awarded the Hughes Medal of the Royal Society for his work on the metal-insulator transition in a wide range of systems and materials. He has been an external advisor to the US Department of Energy research programme and is the UK hydrogen storage representative for the International Partnership for Hydrogen Economy (IPHE). He is also the Executive Director of the UK Sustainable Hydrogen Energy Consortium (UK-SHEC). He has broad research interests encompassing the electronic structure of solutions of alkali metals in non-aqueous solvents, high-temperature superconductors, metal nanoparticles, materials for hydrogen storage, and transparent conducting oxides. He has published over 300 scientific papers and two books.

Professor William I. F. David is CCLRC Senior Fellow based at the ISIS Facility at the Rutherford Appleton Laboratory and Associate Director of Research

Networks at CCLRC. He is a Visiting Professor in Inorganic Chemistry at the University of Oxford. He has been one of the key scientists in UK neutron scattering over the past twenty years playing a pivotal role in the development of the ISIS spallation neutron source. His work has principally focused on using neutron and X-ray powder diffraction to characterize a broad range of new materials. He was made significant contributions to the structural analysis of lithium battery materials, high temperature superconductivity, and C60. More recently, he has transformed the capability of determining organic and pharmaceutical crystal structures from powder diffraction data using global optimization methods. He is currently developing joint X-ray and neutron powder diffraction techniques for the discovery and characterization of novel hydrogen storage materials. He was the recipient of the 1990 Charles Vernon Boys Prize of the Institute of Physics for outstanding contributions to solid state physics and, in particular, to high resolution powder diffraction studies and received the inaugural BCA Prize in 2002 for his outstanding contributions to neutron scattering and powder diffraction. He is the author of over 150 research papers and one book.

Dr Vladimir L. Kuznetsov is a Research Fellow at the Inorganic Chemistry Laboratory, University of Oxford. He obtained his MSc and PhD degrees from Moscow State University (Russia) and worked as a Research Associate at the Department of Chemistry at Moscow State University and then as a Senior Scientists at A.F. Ioffe Physical-Technical Institute, St.Petersburg, Russia. From 1996 to 2004 he was a Research/Senior Research Associate at the Department of Electronic Engineering at Cardiff University. His main research interest is in the area of functional materials for electronic and energy conversion applications, phase equilibria and thermodynamic properties of multicomponent systems and hydrogen storage materials. He has authored and co-authored over 40 journal papers and 6 book chapters.

11. *Fuel cells*

David Jollie

Introduction

The vision of a world without oil or other fossil fuels is both surreal and at the same time seductive as a solution to current concerns over climate change and oil availability. It is also, to some extents, an irrelevant one for fuel cells. Rather than being an energy source they provide a mechanism for transforming one form of energy (chemical) to another (typically electricity or heat). In this way, they resemble batteries, internal combustion engines, and even steam engines. The key to their value is really their efficiency: they are able to carry out this transformation cleanly and efficiently.

Fuel cells are not yet fully developed. The technology and the fuel cell effect were discovered in 1839 by, depending on your point of view, William Grove or Christian Schoenbein (Sanstede *et al.*, 2003). For a long time after this, the technology was essentially dormant until the 1940s when Francis Bacon started working on it and the 1950s when Allis-Chalmers built the first application of the technology (a fuel cell powered tractor). Research and development accelerated when fuel cells were chosen as power sources for space missions in the 1960s and the 1970s oil price shocks increased interest in other technologies, but the real impetus came in the 1990s when DaimlerChrysler examined the proton exchange membrane fuel cell and decided that it could be used to power a vehicle.

Considerable effort is still to be expended on improving fuel cell technology in terms of cost and performance. Ancillary questions like the best method of fuelling and of carrying fuel still remain to be solved. However, we have begun to see fuel cells entering the commercial marketplace and the coming years and decades should see this accelerate.

The principles

A simple definition of a fuel cell might be 'a device that reacts a fuel and an oxidant, without combustion, producing heat and electricity'. The best-known case, that of a proton exchange membrane (PEM) fuel cell (PEMFC), is illustrated in Fig. 11.1.

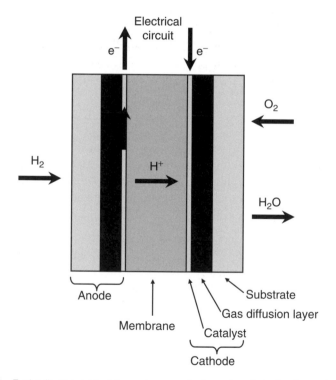

Figure 11.1 Fuel cells convert fuel (here shown as hydrogen) and air directly to electricity, heat, and water in an electrochemical process, as illustrated schematically above for a Proton-Exchange Membrane (PEM) fuel cell. A catalyst in the anode dissociates hydrogen to yield protons and electrons. The electrolyte membrane in the centre of the diagram enables the transport of the protons to the cathode, while the electrons flow through an external circuit to provide electrical current before they reach the cathode. At the cathode, another catalyst assists the combination of the incoming electrons and protons with molecular oxygen from the air to produce water. Many fuel cells can be connected together to provide necessary power.

In a PEM fuel cell, the fuel is hydrogen, the oxidant is oxygen and the only chemical product is water, as described in reaction (1):

$$2H_2 + O_2 \Rightarrow 2H_2O + heat + electricity \tag{11.1}$$

The membrane separates the anode (H_2 oxidation) and cathode (O_2 reduction) but allows the flow of protons produced at the anode and required at the cathode. Like a battery, a fuel cell converts chemical energy into electrical energy, but there is an important difference. A battery typically has a limited amount of energy contained whereas a fuel cell should continue producing power as long as externally-stored fuel and oxidant are fed to it. The electrodes in a fuel cell are inert and 'recharging' is simply a case of adding more fuel rather than forcing the reverse reaction (Hallmark *et al.*, 2004) .

When compared to the internal combustion engine, the situation is quite different. The fuel cell is freed from the tyranny of the Carnot cycle which imposes a maximum efficiency on such engines (typically just under 60% for a realistic diesel engine). No matter what improvements are made to the internal combustion engine (and there are many still to be made), the fuel cell has a higher theoretical efficiency and therefore greater potential. It also has other benefits: where the engine burns fuel at high temperature, producing oxides of nitrogen no matter what, as well as other regulated emissions (carbon monoxide, particulate matter, un-burnt hydrocarbons, and the greenhouse gas, carbon dioxide), the fuel cell reaction does not involve combustion or high temperatures and is correspondingly clean. In terms of environmental and health effects, it is clearly a desirable technology.

Most importantly, the fuel cell is a very flexible device. Its prime purpose is to be a source of electricity (derived from chemical energy) and it can provide this at any scale, from below the few Watts needed by consumer electronics to the Megawatts a power station typically produces.

Of course, the picture is not quite so simple. The fuel cell is really a range of technologies, all operating on the same basic principle, as seen below in Table 11.1. To assess the potential for each fuel cell type in every end use is not within the scope of this article. Here, therefore, the focus will remain unashamedly on the PEM fuel cell, due to its flexibility.

Applications of fuel cells

The fuel cell is widely described as a disruptive technology and this is, in fact, not an inaccurate description. Looking at the battery market today, fuel cells could take market share, making them disruptive. However, changing consumer behaviour is not trivial. Whether it is the best technology or not, a consumer is used

Table 11.1 The main types of fuel cell and some of their attributes.

Type	Operating temperature (°C)	Electrolyte	Typical power range	Typical applications
Alkaline (AFC)	80	Potassium hydroxide solution	100 W–100 kW	Niche uses
Direct methanol (DMFC)	100	Proton exchange polymer membrane	1 W–1 kW	Portable power
Molten carbonate (MCFC)	550–650	Molten metal carbonate	250 kW–2 MW	Stationary
Phosphoric acid (PAFC)	150	Phosphoric acid	50–250 kW	Stationary
Proton exchange membrane (PEMFC)	100	Proton exchange polymer membrane	100 W–100 kW	Portable, transport
Solid oxide (SOFC)	650–1000	Ceramic metal oxide	1 kW–1 MW	Stationary power

to charging a mobile phone from a socket and is willing to do this. Introducing a fuel cell alongside or instead of the battery will require the user to buy fuel cartridges, a different way of behaving and something that will only be adopted relatively slowly.

Fuel cells do, though, have more potential: they could be an enabling technology as well. Where current technology does not perform well enough to allow a new device or product to be introduced, it is possible that a fuel cell would remove this barrier. In this case, the fuel cell is no longer disruptive, it is instead, essential to the product itself. And, it is likely to be here that we will see some early uses of the technology: powering electronic devices (as will be discussed later) is one obvious area and the military establishment is likely to be the first mass user.

Forecasting the future more closely is difficult: the fuel cell is a highly scaleable technology and there are several different types. We could, in fact, see molten carbonate fuel cells in power stations, solid oxide ones replacing boilers in homes, and direct methanol-powered laptop computers. Where the fuel cell story relates most closely to fossil fuels, in cars and in power generation, energy efficiency, energy security, and greenhouse gas emissions are the most powerful driving forces. Elsewhere, other benefits such as a high energy density may be more important.

Market developments and regulation

Given that fuel cells can have a wide range of uses, market developments are very broad, deriving from legislation, research and development, and so on. The introduction of fuel cells in portable electronic devices will, for instance, be led by the marketing departments of the household name manufacturers and by consumer demand for better performance. When we look more closely at a world without oil, though, the most important sector is transport. It is here that the interplay between the manufacturer, the consumer, and the legislator will be most interesting.

For instance, introduction of environmental technologies into consumer markets can be driven by customer interest, but is more often led by legislation. The first catalytic converters fitted to cars in significant numbers were mandated in the 1970s (Kendall, 2004) in order to reduce stunning levels of air pollution in the Los Angeles basin. There were even restrictions on burning coal in London several hundred years ago for similar reasons. Many years later, the effectiveness of such legislation has been seen and regulators still tighten emissions limits to achieve the same effect.

It is to be expected that any widespread use of fuel cells will be at least partially driven by similar legislation. Most government policy, globally, smiles upon emissions reduction, energy efficiency, and new technology, and fuel cells are able to deliver in each of these areas. They will be particularly welcome in the ground transport sector and in the urban environment. Taken together these cause significant damage to human health through economic activity and the majority of anthropogenic greenhouse gas emissions, and therefore provide the best opportunity to gain from the introduction of new technologies. All forecasts of greenhouse gas emissions over coming years see much of the increase in carbon dioxide production deriving from economic growth in developing countries and from an expansion in air travel. Fuel cells, and potentially a range of other technologies, will be required here too.

The vehicle manufacturers will also play a vital role. Already, almost every major carmaker has a fuel cell research and development programme. Some of these are defensive, in ensuring that the company can continue its business whatever happens, and some are more positive in terms of positioning the company as a leader in environmental and technological thinking.

The introduction of hybrid vehicles by Toyota and Honda is a good example where quick action by these two organizations has forced their competitors to think about following suit. This example is particularly relevant for the introduction of fuel cells as there are similarities: customers are being exposed to the fact that environmental benefits can have a value to them; and some components of fuel cell vehicles (such as motors and electronic controls) are being developed for these hybrids. The introduction of hybrid vehicles, with hundreds of thousands being sold worldwide five years after their launch, not only provides a model for how fuel cell cars could be sold but provides a boost to their development and customer acceptance.

Technological developments

From a starting point of 2005 technology, it is clear that technical improvements must be made to fuel cells in order for them to become a realistic alternative technology for the future. Focusing again on PEMFC technology (although it should be acknowledged that equally important challenges exist, and equally striking improvements are being made, in other types of fuel cell), the key challenges come down to cost, performance, and durability, each of which is being addressed by different means.

The cost issue can be split into two parts: materials and manufacturing. Improvements in performance and greater use of purpose-built components both help here, and in improving performance and durability. Manufacturing costs always decrease as production volumes increase, particularly due to adoption of techniques derived from other industries such as semiconductor manufacture. Materials costs pose a more fundamental problem. A simple examination of today's technology confirms that it will not be possible to sell large numbers of fuel cells without significant cost reduction and materials optimization.

The good news, of course, is that this is already happening: companies are examining metal and graphite bipolar (or flow field) plates to find the lower cost option; considerable effort is being expended in designing new membranes (Wakizoe *et al.*, 2005; Foure *et al.*, 2005; Voss *et al.*, 2005) which are either cheaper in themselves or allow a reduction in material cost elsewhere in the system (for instance by increasing the operating temperature and making the role of the catalyst somewhat simpler). One oft-quoted issue is the cost of the precious metal (primarily platinum) in the system. Here it is appropriate to draw a comparison with the history of the platinum content in vehicle exhaust catalytic converters, where it also plays a similar role (Kendall, 2004). Use of chemical promoters to improve the reaction, using a mixture of metals in the catalyst, intelligent design of the catalyst structure, and improved knowledge and understanding of degradation processes, have all allowed dramatic improvements in platinum utilization to be made over the thirty year lifetime of this technology.

Improving performance is largely a question of electrochemistry. Fig. 11.2 shows an ideal performance curve for a fuel cell running on hydrogen and oxygen, and also the causes of deviation from that for today's technology. For a chemist these are the usual issues of activation of reaction at the electrodes and getting the reactant chemicals to the electrodes in the first place (mass transport).

The first issue is a question of catalysis, in terms of speeding the reaction up at low temperature. Here, improvements are often focused on effective utilization of the catalyst (exposing as much of it as possible to the reactants) and on using the best available materials. Mass transport questions are less that of chemistry

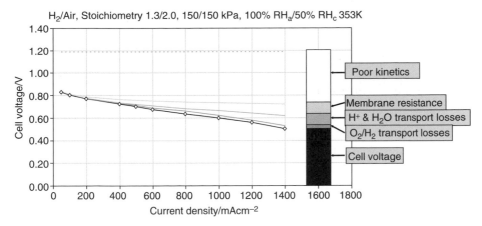

Figure 11.2 Theoretical performance curve for a fuel cell running the hydrogen and oxygen reaction.

and more that of the micro-structure of the fuel cell itself. Here, the structure of the catalyst and how it is applied to the electrolyte or electrode is crucial. For instance, Johnson Matthey has shown the influence of carbon support on platinum surface area and hence activity in direct methanol fuel cells (Gray *et al.*, 2005). The same effects have been seen elsewhere and further progress in terms of improved performance continues to be made.

Improving durability and reliability is perhaps the most empirical of these three areas where improvements must be made. Technologists aim to deal not with an idealized system but with real life. Often components and materials used in prototype fuel cells are not designed for this use, but rather are adopted from other applications. Each of these should be optimized and the problems associated reduced in number and severity. Some issues are basic chemical ones, often due to the accidental introduction of pollutants or the corrosive nature of the electrolyte. Others are more prosaic. Tests of United Technologies' PC25 phosphoric acid units in the field showed that filters clogging and the failure of temperature sensors were amongst the most important failure modes (Binder *et al.*, 2003). A second example from United Technologies focuses on improvements made to its alkaline fuel cells used in the Space Shuttle programme. Here a small design change in terms of the dimensions of the cell components (Poast *et al.*, 2003) improved the stack lifetime significantly, simply by reducing the rate of corrosion of the structural materials by the alkaline electrolyte.

The US Department of Energy has been proactive in setting targets for the performance, cost, and durability that will be required for fuel cells to really bridge the gap between scientific curiosities and practical everyday devices (Adamson 2005). It is not necessary here to list these individual targets and the dates by which they should be achieved, but it is important to note that progress is being made towards meeting them. Industry researchers can see how to meet the earlier goals today, and have indeed already exceeded some of them. The longer term targets by their very nature require more creativity, and mapping the particular technical developments to get there is difficult. The most challenging goals are for the automotive industry, where low cost is of prime importance but cannot be achieved at the expense of performance. The timescale for the widespread introduction of fuel cell cars is therefore correspondingly long, with no serious commentators expecting publicly-available products before 2015 at the earliest. However, there are points along the development curve where other end uses become feasible (for instance the price targets are much less challenging for use in providing domestic power). Any commercial use of fuel cells in these applications will provide a measure of the progress being made towards ensuring that the fuel cell can capture part or even all of the transport market.

From today to tomorrow: changing the game

Although it has already been stated that the fuel cell is an energy conversion device rather than a source of energy, it does fit well with a non-fossil fuel future. Hydrogen is in many ways the ideal fuel for this technology and the combination of hydrogen and fuel cells is powerful, allowing us to conceive of non-polluting energy use. Fuel cells will simply provide the most efficient way of using this hydrogen. (However, it is only fair to point out that technical challenges remain not just with the fuel cell but also with the method of hydrogen storage as mentioned in chapter 10.)

Short term uses of fuel cells may be beneficial in terms of reducing oil and other fossil fuel usage. These are already being seen, in the form of natural gas or biomass fuelled cells (typically either phosphoric acid or molten carbonate cells) providing both heat and power in hospitals and elsewhere. We will see more of this technology being used. Slowly, we are seeing experimentation with other fuels and a move towards sustainability.

The long term aim is universally called 'The Hydrogen Economy', where hydrogen is the dominant energy vector and fuel cells one, or perhaps even the most prominent, energy conversion technology (Adamason, 2005). In this future scenario, oil and other fossils have more value for their chemical properties than simply for their calorific content as fuels, and the value of a fuel cell is as a clean, efficient technology. Renewable energy may be the most attractive energy source in environmental terms, but nuclear power may well also play a part. Already we have seen fuel cells being operated using waste streams from wastewater plants, or in industries such as the chlor-alkali process and brewing. This area will expand, driven primarily by economics.

Fuel cells can also have an important role to play in terms of interaction with renewable power generation. Renewable power is often available only intermittently (most obviously when the wind blows and the sun shines). Conventional electricity grids can only cope happily with a relatively small proportion of such power, limiting the degree of penetration of renewable power generation that can be achieved. Currently it is possible to store this energy (whether on a large scale, say by hydroelectric storage, or on a small scale in batteries) but today's technology is not particularly efficient.

Considerable effort is therefore being expended on the use of electrolysis, hydrogen storage, and fuel cells as a more efficient and scaleable storage method. Remote locations (including islands such as Unst in the UK's Hebrides) could be particular beneficiaries. A visionary view sees the removal of any need for a grid at all, and complete decentralization of power generation using this suite of technologies; although a more pragmatic approach would see hydrogen fuel cells and electrolysis as something that will reduce rather than eliminate the need for a grid. However, the time may come when your whisky is distilled using combined heat and power, derived from fuel cells using hydrogen, formed by electrolysis of water using wave and wind power, on Islay.

Looking further forward into the future, in our oil-less world, fuel cells, presently a disruptive technology, could become a remarkable enabling one elsewhere. At a very practical level, the energy density (amount of energy stored per unit of weight or volume) of a fuel cell and its fuel is much higher than for current or developmental battery technologies. To a casual observer, it seems that the imagination of consumer electronics designers is almost without limit and this group is clamouring for better power sources.

If fuel cells can play this role, we can expect to see ever more complex and capable devices being sold, the shape of which is hard to forecast. The shorthand description is 'convergence devices' where one device adds on the capabilities of another. The increasing functionality of such devices (for instance mobile phones that can show films, take pictures, and send e-mails) requires ever more power. There are doubtless improvements to come in terms of battery technology, but there may be a point beyond which this is not possible. Where this is the case, it is no longer a question of competing for market share: the fuel cell could be the only answer and any such device will use one.

However, there is no doubt that despite the size of the power generation and electronics industries, the real prize for many people remains the introduction of fuel cells into the automotive industry. Here, they could, eventually, replace the internal combustion engine. The challenges to this are clear: a fuel infrastructure needs to be built; costs have to come down dramatically; and, even if it does happen, it will take a long time.

However, the benefits of using fuel cells here are compelling. If clean, sustainable sources of hydrogen are available, then the car ceases to be the environmental monster it currently is. There will clearly still be issues relating to safety and to energy and materials used in construction: although there is no avoiding this, local pollution and greenhouse gas emissions could be, in the long term, drastically reduced or even eliminated. (It is only fair to point out that this must be in the long term: short term sources of hydrogen will very likely still have a carbon footprint, and the overall efficiency of each fuel chain will still be of importance. Analysis of the whole life cycle of a vehicle is important to understand the real environmental impact, rather than simply part of it.)

It is possible, though, to imagine still greater changes to the automotive industry than removal of one of its basic blocks, the engine. Designers at many of the automobile manufacturers have seen that the introduction of the fuel cell and electric drive provides opportunities as well. General Motors' well-publicized 'skateboard chassis' (English, 2005) contains all of the drivetrain for a vehicle within a radically different design to today's vehicles. With motors on each wheel and the fuel cell 'engine' and fuel tank in this structure, the designers already have four wheel drive and are now free to build any cab design on top. Early designs, like Toyota's Fine-S and GM's AUTOnomy (Fig. 11.3) and Hy-Wire seen below, have a little of the 1960s version of the future in them, but refinements like the Sequel, also from General Motors, have already begun to form these into more recognizable shapes. The fuel cell can help to redesign the car, releasing it from the constraints imposed by a fairly solid engine, driveshaft, and other components that control it today.

Amory Lovins, founder of the Rocky Mountain Institute, has moved even further down the conceptual curve. His Hypercar (www.hypercar.com) has all manner of improvements on current technology: not only is the engine replaced by a fuel cell, but the whole car's design is changed to improve the aerodynamics and reduce the weight. The most remarkable part is not what the car is though, but what it can do. Lovins noticed that, although a typical engine might be able to provide 75 or 100kW of power, it rarely does. Even when moving, power requirements are much lower, but the key point is that cars are not in use much of the time at all. A quick calculation shows that, in fact, all of a country's power needs can be met by generating power from a reasonable number of stationary vehicles. The flexibility of the fuel cell, it turns out, means it is ideally suited to playing both roles: powering a car and a house at separate times. This scenario may be quite a stretch even for the imaginative amongst us, but it does illustrate the potential power of this technology.

Figure 11.3   General Motors AUTOnomy concept vehicle shows the design possibilities that introduction of fuel cells allows.

Concluding remarks

The fuel cell itself is not the single answer to a world without oil or hydrocarbons but it does have many useful attributes. It is a cleaner technology than today's alternatives, a benefit for the urban environment in particular. It is also efficient, and can use a variety of fuels, providing more useful power than we get today from combustion engines. The end-point in technological terms is hard to forecast but there is clear potential here.

Industry will continue to develop the fuel cell and the related technologies that will be required, and government support for this is growing as the potential benefits are increasingly recognized. Together with other technologies for power generation, perhaps in the forms of biomass, other renewables, or nuclear power, fuel cells may be able to realize these.

Resources and further information

There are large amounts of published information on fuel cells, often linked to the much-vaunted 'hydrogen economy'. A good starting point is the Fuel Cell Today website, www.fuelcelltoday.com which contains much information on the various relevant technologies and their real-world applications as well as links to other useful information.

To understand the basic technology, a good source is *Fuel Cell Systems Explained*, J. Larminie and A. Dicks, published by Wiley. Considerably greater detail can be found in the *Handbook of Fuel Cells*, W. Vielstich, A. Lamm and H. Gasteiger, also published by Wiley.

Adamson, K. A. 2005. US hydrogen and fuel cell R&D targets and 2005 funding, *Fuel Cell Today* (www.fuelcelltoday.com), January 2005.

Binder, M., Holcomb, F., and Josefik, N. 2003. The DoD ERDC-CERL fuel cell demonstration program, *Eighth Grove Fuel Cell Symposium, September 2003.*

English, A. 2005. *A vision of the future*, The Daily Telegraph, 22nd January, 2005.

Foure, M. *et al.*, 2005. Development of a low cost, durable membrane and membrane electrode assembly for stationary and mobile applications, *2005 Fuel Cell Seminar Abstracts*, 176–79.

Gray, P., Hogarth, M. P., and Cabello, N. 2005. Advances in DMFC catalysts and MEAs for portable and consumer electronics, *Ninth Grove Fuel Cell Symposium*, October 2005.

Hallmark, J., Bostic, E., Jones, H., Burke, T., Kelty, K., Lowe, B., and Wicelinski, S. 2004. *Functional Differences between micro fuel cells and battery packs integrated in portable electronic products*, Underwriters Laboratories, August 2004, UL2265-WD15 Micro fuel cell white paper, http://www.ul.com/dge/fuelcells/whitepaper.pdf

Kendall, T. 2004. *30 Years in the Development of Autocatalysts, Platinum 2004*, 32–37. Johnson Matthey, London .

Poast, K., Burghhardt, M., and De Ronck, H. 2003. Space Shuttle fuel cell life extension program, *2003 Fuel Cell Seminar Abstracts*, 1028–31.

Sandstede, G., Cairns, E. J., Bagotsky, V. S., and Wiesener, K. 2003. History of low temperature fuels. In *Handbook of Fuel Cells*, Vol. 1, (Eds.) Vielstich, W., Lamm, A., and .Gasteiger, H. A. ,p. 156, Wiley, Chichester.

Sorensen, B. *et al.*, 2004. Hydrogen as an energy carrier: scenarios for future use of hydrogen in the Danish energy system, *Intl. J. Hydrogen Energy*, **29**, 23–32.

Voss, H. H., Cha, S., and Chen, E. 2005. Polyfuel's hydrocarbon membrane for transportation applications, *2005 Fuel Cell Seminar Abstracts*, 192–93.

Wakizoe, M., Miyake, N., and Honda, E. 2005. Asahi kasei high temperature membrane for PEFCs with high durability, *2005 Fuel Cell Seminar Abstracts*, 172–75.

The author

Dr. David Jollie is Publications Manager at Johnson Matthey and has worked extensively on hydrogen, fuel cell, and low carbon issues. Trained as a chemist, he moved into analysis of the automotive and fuel cell industries before setting up Fuel Cell Today, a web-based organization aiming to improve the understanding of fuel cells and hydrogen. He has published numerous works on fuel cells and spoken widely on the technology and its competitors.

12. *Energy efficiency in the design of buildings*

Gerhard Dell, Christiane Egger

Introduction

The buildings sector accounts for 40% of European energy requirements. Two thirds of the energy used in European buildings is consumed by private households, and their consumption is growing every year as rising living standards lead to an increased use of air conditioning and heating systems.

Research shows that more than one-fifth of the present energy consumption and up to 30–45 million tonnes of CO_2 per year could be saved by 2010 by applying more ambitious standards both to new and refurbished buildings–these savings would represent a considerable contribution to meeting the European Kyoto targets (European Council, 2002). Without comprehensive measures, energy consumption and CO_2 emissions from the building sector will continue to grow.

Sustainable energy strategies for buildings will therefore increase in importance. Even today, so-called 'zero emission buildings' can be realized with existing planning approaches and technologies. Such buildings do not need an external energy input (for example from oil, gas or supplied electricity) other than solar energy. This is achieved by a combination of a high-level of energy efficiency and renewable energy technologies. This chapter focuses on buildings in the housing and service sectors, presents new building design strategies, technologies, and building components as well as the new legal framework set by the European Buildings Directive. It also discusses the question of raising awareness, and presents some thoughts on how changing life patterns may impact the buildings of the future.

Residential buildings mainly need energy for space heating; with present building standards, space heating represents about 70% of the overall energy demand of existing buildings. In many European countries there are substantial efforts to increase energy efficiency—nevertheless, not all the potential for energy savings has been realized by far, and oil is still a major energy source for heating. In recent years, heat demand for new buildings was reduced significantly by technical measures. However, the number of low energy or passive buildings in Europe is still very limited, despite the fact that they can be constructed at acceptable costs.

In addition to the reduction of the energy demand of existing residential buildings, new building design strategies must show the way towards significant decreases in building energy demand. Office buildings usually have higher electricity consumption, but lower heat demand, than residential buildings.

The increase of energy efficiency must take place throughout the entire energy conversion chain (see Fig. 12.1), from the provision of energy services to the delivery of primary energy; and in the future primary energy should be from renewable resources which do not exacerbate climate change.

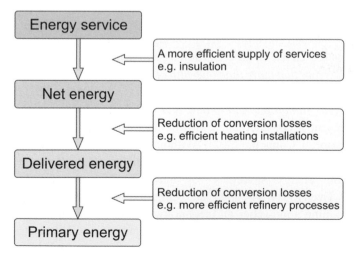

Figure 12.1   Illustrates the principle steps of energy transformation.

Low-energy buildings

In order to be able to compare the thermal quality of buildings and to specify requirements for new construction or renovation, so-called 'energy performance indicators'—documented in an energy certificate—are increasingly used throughout Europe. Similar to the fuel consumption per 100 kilometres, which shows how economical or wasteful a car is, the energy performance indicator expresses—for example—the annual heating energy demand of a building per square metre.

Amongst other factors, the energy performance indicator is determined by the building design, the orientation of the building, and solar gains thorough windows as well as the insulation characteristics of the different building components. Additionally, the HVAC (heating, ventilation, air-conditioning) systems play a crucial role, for example, in the influence on the energy performance indicator of the use of mechanical ventilation systems with heat recovery. In the energy flow diagram shown in Fig. 12.2, the energy demand up to primary energy sources and the respective CO_2-emissions can be seen.

Nowadays, buildings are defined as 'low-energy buildings' if their heat demand is a third below the legally-defined minimum standards. 'Passive buildings' are a further development of low-energy buildings. In a passive building, the heat losses through the building shell and through building ventilation are so reduced that the solar gains through the windows are sufficient to limit the annual heat energy demand to 15 kWh per m^2 living area (in Central European climates). This is a reduction of more than 80% compared to the legally-required energy efficiency standards. This can only be achieved by very high insulation standards, by high quality of workmanship, and by mechanical ventilation with heat recovery. These aspects have significant implications for the way buildings are designed.

So-called 'solar low-energy buildings' do not only rely on energy efficiency. By using small pellet stoves for space heating, as well as solar thermal collectors for domestic hot water and for seasonal heating, a high percentage of the energy demand can be covered by renewable energies. The example below, (Fig. 12.3) from a region of Upper Austria shows how this can be done in practice even today.

In a region of Upper Austria, a comprehensive building programme has cut energy consumption in 95% of all new homes by 50% since 1993. This was achieved through a soft loan programme which combines a financial incentive with targeted information. The calculation of an energy performance indicator, and the participation in an obligatory individual energy advice session of the homeowners and an energy performance certificate for every building, are the most important programme elements. If the requirements are met, the home owners receive an additional low-interest loan (Dell, 2001).

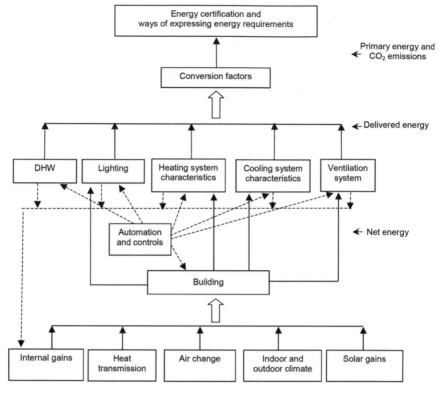

Figure 12.2   Diagram of the building energy flow. Source: CEN.

Figure 12.3   Solar low-energy building in Upper Austria.

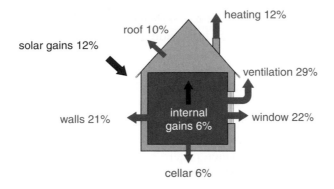

(for new and retrofitted homes)

Figure 12.4   Distribution of heat losses in a home.

From 1993 to 2005, more than 50,000 homes met the programme requirements, achieving an energy saving of 350 million kWh/year. This leads to a CO_2 reduction of 70 million kg/year. The programme is also very attractive in economic terms: using the least-cost-planning calculation method, every kWh saved through energy efficiency measures costs only 1.8 Eurocent.

Technologies

Energy losses of buildings occur through all external construction components, mostly through windows and external walls (see Fig. 12.4). Good insulation characteristics of the construction materials and components (expressed in 'U-values[1]'), especially those surrounding unheated space and the outside, lead to a decrease of the energy demand.

Good levels of thermal insulation are, for example, windows with U-values between 1.1 to 0.5 W/m^2K; attic floors with a depth of 25–40 cm insulation material; external walls, for example, 38–50 cm wide; and insulation bricks or double walls with 12–20 cm insulation between or wood latch plate wall with 20–30 cm of insulating material.

High-efficiency windows—optimized both for insulation requirements and for solar gains—play a crucial role in the market development of low energy and passive buildings. Therefore, new mass market windows were developed in recent years, which—by applying special coatings—combine a high transparency

---

[1] U-Value: The heat transfer co-efficient of a material or an assembly of materials. The lower the U-value, the better (the greater the heat transfer resistance (= insulating) characteristics of the material or assembly of materials).

for the whole of the solar spectrum with low heat losses within the infrared range. Cost-efficient anti-reflection coatings can lead to further technological improvements.

The specific heat losses of opaque walls of low energy/passive buildings should be below 0.15 W/m² K. Using standard materials, this can be achieved relatively easily with diameters of 0.3–0.4 metres. However, thick walls can cause aesthetic problems, especially for the integration of windows. A new building technology, so-called 'vacuum insulation panel' have great innovation potential in the near future. With evacuated insulation panels, heat conductivity is up to 5–10 times lower, compared with conventional insulating materials. In flat panels—different from round containers, such as thermos flasks—the load of the outside air pressure of 1 bar is transferred to the padding (Bayerisches Zentrum, 2006).

Transparent insulation materials make use of passive solar gains. An example is the so-called 'solar facade' (see Fig. 12.5 and Fig 12.6.). This facade system is made up of a special cellulose 'comb' which is placed under a glass panel on the outside of external walls. In wintertime, the light of the low winter sun penetrates the 'solar comb' and heats it up. Thereby, a warm zone is formed at the exterior of the wall. In summertime, due to the higher position of the Sun, the structure of the solar combs casts a shadow, and thereby eliminates overheating. The effectiveness of the solar facade depends on the supply of sunlight. South, east, and west sides of the building are suitable for solar facades. Here, walls with average U-values in the range of 0.05 W/m²K are possible—which means that heat losses are practically non-existent.

Covering the minimal heat demand of low energy and passive buildings is a new challenge for HVAC engineering. The smallest-scale biomass heating systems (a few kW installed capacity) are already available on the market (see Fig. 12.7). Furthermore, the integration of small-scale fuel cells into building services concepts is also being discussed for the long term.

Such a fuel cell, operated by solar-generated hydrogen, can be integrated as a module into a mechanical ventilation system instead of heat pumps which are frequently used at present. Both the supply air and the hot water can be heated by the waste heat of the electricity production. The best efficiency of fuel cells can be achieved if electricity and heat requirements are similar. Fuel cells operated by solar-generated hydrogen are complementary to other decentralized electricity generation systems, such as wind, PV, biomass, and biogas.

Heating with solar collectors is already state-of-the-art today, although it remains costly with a pay-back time of ten years. However, if high solar coverage is to be achieved, large storage tanks are necessary, even for small heat demands (see Fig. 12.8). The on-going development of thermochemical heat storage systems based on absorption and adsorption processes has offered new impulses over the past years.

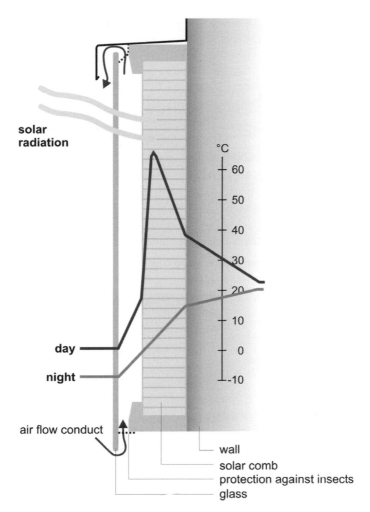

Figure 12.5 Function of transparent insulation—'solar façade'—system. Source: Gap-solar (GmbH).

Such storage systems consist of containers with silica-gel as reversible adsorbent for water vapour. The research target is to develop storage densities of the order of 200 kWh/m², which work practically without storage losses. Materials research efforts will lead to innovations in this field in the near future. Such storage materials would exceed the energy density of conventional hot water storage tanks by the factor 4. With high storage densities of future systems, higher solar coverage rates can be achieved.

An example of successful development in fuel switching is that which took place in Upper Austria in recent years (see Fig. 12.9). The percentage of oil

Figure 12.6   Example of a building with transparent insulation in Upper Austria.

Figure 12.7   Example of a small-scale biomass heating system. Source: Rika GmbH & Co KG.

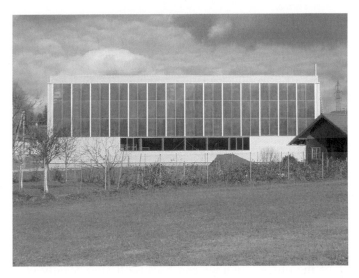

Figure 12.8 Example of a solar thermal collector system at a production hall in Tumeltsham, Upper Austria.

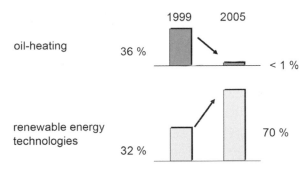

Figure 12.9 Fuel switch in new homes in Upper Austria.

heating systems in new homes decreased from 36% in the year 1999, to under 1% in 2005, and in the same period the percentage of renewable energy heating systems increased from 32% to 70%.

Europeans spend 90% of their time in buildings, either at home or at work. Therefore, it is worthwhile choosing building materials carefully. Insulation materials like Polystyrole, which are used worldwide, are an oil refinery product and cause styrene and pentane emissions at production. Today, these insulating materials could simply be replaced by fossil-fuel-free materials, such as cellulose, sheep wool, or cork.

Tertiary buildings

At present, modern office buildings have a typical energy demand of approximately 200 kWh per m² per year. 50% of this is electricity (Luther, 2001). The main reasons for this high percentage (compared to residential buildings) are—apart from the technical equipment—the higher density of people, as well as higher requirements regarding lighting and indoor climate. Different from the domestic sector, electricity consumption is largely determined by the technical infrastructure of the building—and not only by the equipment used (such as computers, printers etc.). Today, solar designs for office buildings include an improved use of daylight and the replacement of active air conditioning systems by so-called 'passive cooling'. Modern building design successfully integrates all relevant factors, including lighting conditions, as well as protection against summer overheating (see Fig. 12.10).

Daylight should replace artificial light wherever possible, both for the benefit of the users of the office buildings and because it significantly contributes to energy efficiency. An office without daylight has an electricity demand of at least 20 kWh per m² per year (assuming a daily use of about 8 hours). Intelligent planning of office lighting also avoids summer overheating, as daylight has a higher luminous efficiency than the usual artificial lighting systems and decreases cooling loads. In terms of 'visual comfort' artificial light is significantly worse than daylight. Although glare-free lighting of workstations can be obtained more easily using artificial lighting, the visual contact to the outside as

Figure 12.10 A solar low-energy service building in Upper Austria.

well as the dynamic nature of daylight are important aspects ensuring visual comfort.

Integrating the use of daylight into building design has to start with the architectural draft, and also includes the selection of the suitable, complementary artificial light system. Materials research activities offer many possibilities for the improvement of the use of daylight.

Creating an agreeable indoor climate—especially in summer—must be an important objective of any good building design. With the currently prevailing 'glass architecture', this is a difficult task, especially if an active air conditioning system is to be avoided. Reasons for not wanting to install such a system can be economic, or fear of the so-called 'Sick-Building-Syndrome'. Increasingly, building designers are requested to construct buildings that achieve a good indoor climate in summer and winter without an air conditioning system.

Passive cooling systems function without compressors but instead use natural cooling sources, like the soil, the groundwater, or cool night air. Experience shows that even large office buildings can be cooled effectively this way (Dell & Probost, 2006). It is, however, essential, to take this into consideration at the very early stages of building design—just as in the case of daylight.

The thermal storage capacity of the building itself is an important factor for the effectiveness of passive cooling systems. In addition to using heavy building materials and ensuring a good thermal contact of the construction elements to the indoor air, a further decrease of summer peak temperatures can be obtained with latent heat storage materials.

Seasonally, and often also during the daily cycle, high cooling loads and solar gains coincide. Therefore, solar cooling offers direct use of solar energy for cooling without complex storage systems. Solar thermal collectors work very effectively in thermodynamic installations, which can be operated at low temperatures. Different processes can be used (Richler 2005): about 60% of all solar cooling installations in Europe use 'closed absorption', about 12% an 'adsorption' system. The remaining plants (28%) cool the supply air of the building directly, with so-called 'open sorption processes'.[2]

---

[2] The basic principle behind (solar-) thermal driven cooling is the thermo-chemical process of sorption: a liquid or gaseous substance is either attached to a solid, porous material (adsorption) or is taken in by a liquid or solid material (absorption). The sorbent (e.g. 'silica gel', a substance with a large 'inner surface area') is heated (e.g. by a solar collector) and is thereby dehumidified. After this 'drying', or desorption, the process can be repeated in the opposite direction. When steam is added, the heat is stored in the porous storage medium (adsorption) and simultaneously heat is released. There are several processes using different materials and systems.

The new directive on the energy performance of buildings

With 40% of the EU's energy consumption, the buildings sector offers the largest single potential for energy efficiency. Energy efficiency potentials can be found in many areas: ten million boilers in European homes are more than 20 years old; their replacement would save 5% of energy used for heating. Around 30–50% of lighting energy could be saved in offices, commercial buildings, and leisure facilities by using the most efficient systems and technologies. Half of the projected increase in energy needed for air conditioning—expected to double by 2020—could be saved through higher standards for equipment, as well as different techniques.

The aim of improved energy efficiency in buildings has been set out in existing legislation, in many European countries and regions as well as on a European level. The Directive on the energy performance of buildings (European Council, 2002), in force since 2003, builds on those measures with the aim of significantly increasing the energy performance of public, commercial, and private buildings in all member states. The Directive is a vital component of the EU's strategy to meet its Kyoto Protocol commitments. Meeting the commitments will strongly depend on how well the Directive is implemented.

The Directive—which member states need to incorporate into national legislation by 2006—will ensure that building standards across Europe place a high emphasis on minimizing energy consumption. This will reduce the use of energy in buildings across Europe, without requiring huge additional expenditure, whilst at the same time perceptibly increasing comfort for users.

Under this legislation a common methodology for calculating the energy performance of a building, taking account of local climatic conditions, will be applied throughout the EU; minimum standards for energy performance will be determined by member states, and applied both to new buildings and to major refurbishments of existing large buildings. A system of building certification will make energy consumption levels much more visible to owners, tenants, and users; boilers and air conditioning systems above minimum sizes will be inspected regularly to verify their energy efficiency and greenhouse gas emissions.

Energy Awareness

Looking solely at energy balances and the technological solutions for a building is certainly insufficient, the needs of human beings, the building users, have to be taken into consideration as well.

It may sound trivial but is often forgotten in practice: energy efficiency is only possible if the private household, the company, or the public body is also able and willing to implement appropriate actions and if sufficient capital for investments or new financing instruments (e.g. energy performance) contracting are available. In Upper Austria, a high awareness level for energy topics has been achieved through targeted information activities over many years (Egger *et al.*, 2000). An important aspect is the fact that people have to be aware of how they can save energy or use renewable energy sources in their own sphere of influence. With regard to planning activities, the knowledge of availability is necessary for future effect (Watzlawick, 1995).

Experience shows that higher expectation levels regarding energy efficiency and renewable energy sources bring better effects—with an appropriate feedback loop—than a goal that is set too low. Knowledge of the possibility of success creates an attitude which contributes substantially to achieving it (Bergus, 1984).

One of the problems of energy-related decisions (e.g. the selection of a heating system) is that the consequences of these decisions partly have long term effects which are beyond individuals' perspectives. Therefore, the finiteness of certain energy sources or the long term risks of certain energy technologies, are only partly taken into consideration when it comes to individual decisions.

The exposition of facts, as it is done in energy programmes and strategies, is necessary as a decision base for expert; however, they have often little emotional value. Only personal experiences; negative ones like the smoking chimney of a factory, country-wide floods, or the positive experience of hot water directly from the solar collector, induce emotions which are necessary for a change of behaviour. This can be supported by the fact that important persons (e.g. relatives, the 'boardroom', one's own children, 'authorities') have the same attitude. This social factor also includes the willingness to accept and obey regulations as well as advice from experts.

Appropriate information must be available at the right time and at the right place. Therefore, the questions to be answered are:

• Under which conditions do individuals and enterprises make their plans for the future?

• Which type of knowledge has the strongest impact on decisions?

The O.Oe. Energiesparverband was established at the beginning of the 1990s by the regional government of Upper Austria, to provide comprehensive energy information and advice to all energy consumer groups. The main targets of this regional energy agency with international tasks are: the promotion of efficient energy use, and the increased use of renewable energy sources and of innovative energy technologies. The O.Oe. Energiesparverband is a central contact point for energy information, technology development, and energy advice as well as a contact address for financial support programmes and energy questions in Upper

Austria. With 15,000 energy advice sessions held every year, it is one of the largest providers of energy advice in Europe. Energy advice is offered not only to individuals, but also to all companies and public bodies in the region.

The future of building and living

The conditions for the buildings of tomorrow are influenced by the following presumptions (Haus der Zukunft, 2006):

- we will live longer and remain a lot more active;
- we will often live between the city and the countryside;
- we live alone or with a second person;
- the principle will continue to apply: born to move;
- we work differently and longer;
- and—*last but not least—an increased use of energy efficiency technologies will take place.*

The living practices of tomorrow will become a multi-coloured mirror image of the various, individual life styles. Flexibility will become more important, especially in our homes. Preferred dwellings will be those that are adaptable to changing life styles of their inhabitants. Quality of life will increasingly be defined by the quality of our homes. This includes access to free spaces as well as to leisure facilities.

Summary and conclusions

The building sector with its high, mostly unrealized potential for energy efficiency and renewable energy sources, represents a major opportunity for phasing out non-renewable energy supplies which contribute to climate change, and at the same time offers attractive options to meet our climate commitments. Therefore, a consequent implementation of European Buildings Directive is of utmost importance.

The increased insulation of buildings, as well as the use of innovative technologies, do not only bring advantages in increased user comfort but also contribute also to the creation of jobs.

The acceptance of individuals as users of buildings is essential and must be kept in mind. Without any doubt, it will be possible to operate all buildings without fossil fuels, within just a few years.

References

Bayerisches Zentrum für Angewandte Energieforschung e.V. Abteilung 2, viewed 06 January, 2006, http://www.zae-bayern.de/deutsch/abteilung-2/projekte/archiv/vip/.

Bergus, R. 1984. Psychologische Paradigmen und theoretische Ansätze in Forschungen zum Energiesparverhalten und zur Energiepolitik. *Psychologische Beiträge,* Band 26, 167–84.

CEN umbrella document CEN/BT WG 173 EPBD N 15 rev.

COM, 2003, *Explanatory memorandum to the proposed directive on energy end-use efficiency and energy services*, 739.

Dell, G. 2001. *Bauförderung-Von Niedrig-bis Nullenergie*. Alpenreport. Verlag Paul Haupt, Bern.

Dell, G., Probost, J. 2006. *Sommertauglich entwerfen und bauen*, O.Ö. Energiesparverband/Bremer-Energie-Konsens GmbH, Linz.

Directive 2002/91/EC of the European Parliament and of the Council of 16th December 2002 on the energy performance of buildings.

Egger, Ch., Öhlinger, Ch., and Dell, G. 2000. Making things happen! *Renewable Energy World*, **3**, 4/July–Aug, James & James, London, http://www.gap-solar.at.

Haus der Zukunft, http://www.iswb.at/openspace/gebaut2020/index.htm.

Luther, J., Wittwer, V., und Voss, K. 2001. Energie für Gebäude – solare Technologien und Konzepte, *Physikalische Blätter* **57** Nr. 11, 39–44.

Richler, N., 2005. *Climasol, Leitfaden zum solaren Kühlen,* Rhonalpenergie/ O.Ö. Energiesparverband, Lyon/Linz.

RIKA G.m.b.H. & Co KG, A-4563 Micheldorf, viewed 06 January, 2006, http://www.rika.at/de/2143/.

Watzlawick, P. 1995. *Wie wirklich ist die Wirklichkeit? Wahn, Täuschung, Verstehen.* Serie Piper, Munchen.

The authors

Dr. Gerhard Dell is the energy adviser of the regional government of Upper Austria and the Managing Director of O.Ö. Energiesparverband (ESV), the energy agency of Upper Austria. ESV is an organization set up by the regional government to promote energy efficiency and renewable energy sources. Trained as an electrical engineer and energy economist, he has been responsible for designing and implementing a number of successful legislative and financial initiatives in the field of sustainable energy, such as the 'Sustainable Housing Programme of Upper Austria'. He is a lecturer at the University of Linz and received the Austrian National Award for Energy Research.

Christiane Egger is the deputy Managing Director of O.Ö. Energiesparverband, the energy agency of Upper Austria. She has university degrees in law as well as in environmental engineering. She is the Vice-President of the advisory committee of the European Commission in the field of energy and transport and chairs the working group on 'Sustainable Energy Policies'. She is the conference chairperson of the 'World Sustainable Energy Days', one of the largest annual international conferences in this field in Europe. She has published numerous articles on sustainable energy production and use and spoken widely on technology and policy development in this field.

The O.Oe. Energiesparverband is a regional energy agency set up to promote energy efficiency, renewable energy sources and innovative energy technologies located in Linz, Austria. Main target groups are private households, trade/commerce and industry, and professional associations. The O.Oe. Energiesparverband works to promote energy efficiency and the use of renewable energy sources and new energy technologies. The agency's main objective is to help energy consumers use energy more efficiently, thereby reducing the environmental load. The O.Oe. Energiesparverband is a regional energy agency organized as a non-profit association founded by the Upper Austrian government. For more information see www.esv.or.at.

13. *Governing the transition to a new energy economy*

James Meadowcroft

Over the next two or three decades a new energy economy should begin to take shape in the developed industrial countries. This will not be a post-fossil fuel economy. But it could be an economy in which non-fossil sources play a more important role; where efficiency in the production, distribution, and use of energy is significantly enhanced; where new storage and carrier technologies are being adopted; and where the fossil sector is being transformed by the imperative of carbon sequestration. Such an energy economy would represent a critical staging post in a much longer transition towards a carbon neutral, low-environmental impact, energy system. The extent to which a new energy economy actually materializes will depend on many factors including the pace and orientation of international economic development, the rate and direction of technological innovation and diffusion, as well as patterns of geo-strategic cooperation and conflict. But there is no doubt the trajectory will be significantly influenced by political decisions and government action on the energy file. This is the issue with which this chapter is concerned.

At the moment there are two main political drivers for the move to look beyond oil. First, there are supply concerns. Increasing global demand, production bottlenecks, and political instability have pushed oil prices towards historic highs. Although the oil intensity (oil consumption per unit of GDP) of the OECD economies is less than during the oil crises of the 1970s (IMF, 2005), there is no doubt that the long term economic impact of high oil prices would be considerable. There are also critical issues associated with the geographic distribution of reserves. Production from areas opened up following the turbulence of the early 1970s (such as the North Sea) is peaking. In coming years the United States will be more heavily dependent on imported oil, with an increasing percentage of these imports destined to come from politically volatile areas in the Middle East and Asia. And this presents a serious risk of supply disruption. Meanwhile, the debate about the extent of conventional oil reserves simmers in the background, with recognized experts differing over whether the global 'Hubbart Peak' is already upon us, or lies several decades in the future. Second, there are environmental concerns. The oil economy (and fossil fuel usage more generally) is associated with a host of environmental problems including urban smog (ozone, particulates), acid deposition (emission of nitrogen oxides and sulphur oxides), and toxic releases (mercury) (Ristinen and Kraushaar, 1999; EEA, 2002). While developed countries have made some progress in managing these stresses, on a global scale they continue to rise. Above all, climate change looms on the horizon as the most complex and potentially damaging environmental problem with which human kind has had to deal.

Both these drivers are operating, but in an uneven and partly contradictory fashion. The strength of the supply driver has fluctuated with the price of oil, although the seriousness of the strategic problem has now begun to register with political elites in many countries. Indeed, the US president has spoken recently of the need to break his country's 'addiction' to oil, and to reduce dependence on imports from the Middle East. But this driver does not only point 'beyond oil': in the short term it can lead to intensified efforts to expand oil production, secure new supplies, develop unconventional oil resources, and so on. Even when the focus is not oil itself, this driver can motivate an additional commitment to fossil fuels as a way to diversify energy portfolios or reduce foreign dependence—for example, by turning to coal for electricity generation or developing coal-to-oil and coal-to-gas conversions. Over time the environmental driver is strengthening, but it is vulnerable to the business cycle and political contingencies. It is still early days for climate change politics and probably only when adaptation costs soar, and impacts become undeniable, will it become a decisive factor in energy decisions. Of course, this driver points away not just from oil, but from fossil supplies more generally—particularly from coal, the most carbon intensive of modern fuels. Here climate change and other environmental considerations pull in the same direction. But much depends on the technologies associated with particular energy sources and the assessment of the significance of diverse environmental impacts. For example, the nuclear industry has long been a target of environmental campaigners—but if the relative risks of climate change are ranked high, the environmental costs of a new generation of fission reactors might be seen as acceptable in the race to drive down carbon emissions during the decades before other technological options become available.

Evidently we are headed for a world 'beyond oil'. But the timing and the nature of the path into such a world is clouded with uncertainty. Two of the largest uncertainties relate to the true extent of remaining oil resources (and the technological requirements for their eventual extraction), and the 'sensitivity' of the climate system (how serious the problem turns out to be, and how quickly significant impacts become manifest). What is clear is that on a scale of a few decades what lies 'beyond oil' is actually a great deal more oil (and other fossil fuels as well). Today about 35% of global primary energy supply comes from oil, with another 45% derived from other fossil sources (IEA, 2004). Modern transport is almost entirely dependent on oil. More than half the oil used since commercial exploitation began in 1860 has been consumed since 1985 (Boyle, Everett and Ramage, 2003). And the International Energy Agency assumes that world demand will grow by more than 50% over the next twenty years (IEA, 2004). Through exploitation of more remote deposits, enhanced recovery, development of unconventional sources (tar sands and oil shale), deep ocean drilling, and so on, the industry will chase down every potential supply, although at increasing cost. Our technological infrastructure has been designed and built for oil and other fossil fuels. And they will not be abandoned lightly.

Energy politics

For most of the twentieth century the energy sector in developed states was subject to high levels of state control. Energy companies (especially in the electricity sector, but also in coal, oil, and gas) have often been under public ownership. But even where this was not the case governments have attempted to control fuel choices, investment, pricing levels, corporate mergers, and operating practices. National security, economic and industrial policy, equity, and public health and the environment provided ample justification for the establishment of complex systems of vertical and horizontal governance that influence national energy development trajectories. In most contexts the basic belief has been that the stakes in the energy game are just too high to leave things entirely to markets and private operators.

Since the early 1980s there has been a tendency for government to pull back from direct intervention in economic affairs. This has led to de-nationalization and de-regulation in the energy sphere. There has been a rise in cross-national energy flows, and particularly in the electricity sector there have been experiments with privatization and deregulation (Harris, 2002; Plourde, 2005). The impact of these reforms on supplies, prices, and fuel mixes, as well as their political fall-out, has varied from jurisdiction to jurisdiction. Certainly they have altered the context for making energy policy in the developed world (Doern and Gattinger, 2004). And yet, overall, energy remains among the most densely regulated economic sectors.

The everyday business of energy politics is typically conducted out of the public eye, in endless encounters between regulators, politicians, and energy executives. Producer groups—whether based in the public or the private sector—have traditionally dominated these processes. The political pull of the oil and gas industry is legendary, but the history of nuclear power provides another illustration of producer-driven development. It is only when things begin to go badly wrong that energy politics becomes the stuff of headlines. This was the case across the developed world following the oil shocks of the 1970s. But most countries have had episodic crises, accidents, scandals, protests, and political infighting that have temporarily brought energy issues to the foreground. The Enron scandal and the brownouts in California (in the early 2000s) that accompanied partial electricity deregulation provide perfect cases in point. In many countries the debate about nuclear power has never really gone away. And arguments about other environmental pressures have been ongoing—acid rain, and more recently climate change. But after oil supplies and prices settled down in the early 1980s, energy remained of comparatively low political salience for nearly two decades.

Now this looks set to change. Rising prices and pictures of motorists queuing at the fuel pumps mean that the spotlight is back on oil. Political leaders are being pressed for action, and international gatherings such as the G8 are once again preoccupied with energy. Yet any quick solution to current difficulties is doubtful. With demand strong, and refining capacity stretched, supplies remain vulnerable to political or meteorological convulsions. Nevertheless, over time new supplies will be brought forward. Prices can fall as well as rise. And a serious global recession (perhaps partly sparked by high energy costs) could see demand plummet. But even if supply worries abate, pressures from the climate change side of the equation will continue to grow. So it may be that we are at a political turning point—where energy policy can no longer recede into the background. And the recent call by the Swedish Prime Minister to end his country's fossil fuel dependence within 15 years represents one form of response to the new conjuncture (*Sustainable Industries Journal*, 2005).

Sustainable energy policy

Sustainable energy policy provides an appropriate policy frame to approach energy issues 'beyond oil'. Political decision making on energy matters should be related to the broad goal of sustainable development. According to the oft-quoted definition, sustainable development is a process of social advance that 'meets the needs of the present without compromising the ability of future generations to meet their needs' (WCED, 1987). It implies a development trajectory that enhances societal welfare, while paying particular attention to the plight of the world's poor and to respect for environmental limits. It expresses the intuitive idea that in working to improve our own lives we should not neglect those who have the least, and we should avoid 'fouling the pond' for people who come after us. Thus it embodies ideas of inter- and intra-generational justice, while emphasizing the importance of protecting global eco-systems. It is a normative concept—much like 'democracy' or 'freedom'—that reflects widely accepted values (Lafferty, 1996). And while disagreements about its meanings and practical implications are inevitable, it can provide an important grounding for public debate and decision making (Meadowcroft, 1997).

'Sustainable energy policy' is energy policy oriented to contribute towards sustainable development. It is not just about the environment, for it engages with energy in relation to the overall welfare of societies. Nor is it just concerned with 'renewables'—energy systems that can in principle operate indefinitely because they harness recurrent natural flows such as sunshine or wind. Such alternatives already make a contribution, and their dramatic expansion will be critical to the emergence of a carbon-neutral, and low environmental impact, energy economy (Boyle, 2003). But fossil fuels underpin present livelihoods, and they will continue to dominate global energy supply for decades to come. So sustainable energy policy must also be concerned with non-renewables, with how they can be used effectively and in the least damaging fashion. Issues of energy efficiency, energy conservation, and demand reduction are therefore central. So too are techniques to minimize the environmental impacts of non-renewables. Carbon sequestration may prove critical in this regard, as are measures to reduce other environmental burdens imposed by fossil fuels and nuclear power. Above all, sustainable energy policy is concerned with orienting action to meet current economic and social needs while accelerating the transition towards a carbon-neutral, low-environmental-impact, energy future. So it is a perspective that sets current decisions within the framework of a long term transformation of the energy system (Doern, 2005). Key considerations for such an orientation include:

- Integrating economic, social, and environmental dimensions in decision making. Energy issues should be approached 'in the round', with their potential impacts on economy, equity and environment kept in focus.

- Strengthening the resilience of energy systems. Traditional problems of 'security of supply' are important, but so too are broader questions about the development of energy infrastructure in the face of uncertainty, managing risks, and avoiding premature technological 'lock-in' (difficult to reverse commitments to a technology that may later prove to be less attractive than alternatives).

• Incorporating an 'internationalist' perspective. It is not just that energy markets are increasingly international, and that events abroad can affect domestic supplies and process. It is also that global economic and environmental issues are increasingly inter-twined. So, for example, 'energy assistance' (including technology transfer) to developing countries may become an essential strategy for domestic (and global) environmental protection.

Socio-technological transitions

The energy economy involves an array of more or less tightly coupled socio-technical systems, concerned with various aspects of the production, transformation, distribution, and consumption of energy. These systems involve complexes of interdependent technologies that cross-link to other parts of the energy sector and outwards to the wider economy. Technologies are embodied in physical infrastructure, but they also involve interactions among the social organizations that own and operate facilities, provide finance, furnish equipment and services, consume outputs, conduct research and development, train personnel, and regulate the sector (Bijker, Hughes and Pinch, 1989). The oil and gas industry is physically embodied in oil rigs, pipelines, refineries, and retail filling stations. But its technologies and infrastructure are institutionally linked to the operation of business corporations, to markets for stock, futures, and insurance, to bodies charged with regulatory oversight, to the curriculum of engineering schools, and so on. Over time, nested hierarchies of technologies co-evolve, with changes in one system sparking adaptation and adjustment in related fields (Geels, 2002). And these technological developments are bound up with the evolution of the related social organizations (Bijker, 1995; MacKenzie and Wajcman, 1999). Because each technological component is tied to other elements (even if on the macro scale the degree of overall integration among sub-systems remains loose), technological innovation usually proceeds in small steps—with continuous improvements in performance, manufacturing techniques, cost efficiencies, and so on. Larger-scale adjustments, characterized by the wholesale replacement of one technology by another (from gas to electric lighting, for example) are less common, with higher risks but also with potentially larger rewards for innovators.

Historical experience with the transformation of large socio-technological systems suggests a number of lessons that are worth keeping in mind when reflecting on the emergence of a new energy economy. First, technological change is not just about scientific discovery and engineering prowess. It is also about altering patterns of social organization and interaction (Bijker, 1995). Changing technologies means altering established behaviour. All sorts of obstacles—in addition to the purely technical—stand in the way of doing things in a new way. It is not just that an emergent technology must be operable, and be operable in a cost-competitive manner. It must also be cast in a form that is compatible with, or that forces an adjustment to, established business practices, existing policy frameworks, entrenched customer expectations, and dominant social attitudes. Issues as apparently distinct as the functioning of capital markets, the practices of the insurance industry, the operation of regulatory regimes, and the tastes and concerns of consumers can influence the relative success of specific technological ventures. The point here is that if we intend to accelerate technological transitions in the energy sphere we must be at least as concerned with innovation in the business, regulatory, and consumption spheres as with the actual process of scientific and technological discovery.

Second, socio-technological change is characterized by great uncertainty (Berkhout and Gouldson, 2003). It is impossible to know in advance which technologies will prove to be 'winners' and which will result in failure. Encouraging leads may go nowhere, while an area of enquiry that has been stagnant for years may suddenly come to life. Governments do not have a good record of predicting technological trajectories. But even in the private sector, grand technological visions are far more likely to end in fiasco than to be realized in anything like their original form. Change often turns out to be both quicker and slower than expected. The most fruitful application of a discovery may be far from the realm that originally motivated research. There are almost always unforeseen economic, social, and environmental impacts. Perhaps the most important lesson for policy-makers here is to focus on the broad picture, improving general framework conditions for technological advance in a targeted area, rather than trying to micro-manage the innovation process. Thus multiple technological options should be pursued and the early reduction of alternatives should be discouraged. In other words, policy is best directed to functional ends (Kemp and Rotmans, 2003)—based on the identification of societal needs (for example, the need for efficient, cost-effective, low carbon emission fuels), rather than to backing favoured technologies (say those that appear closest to market, or that have strong national champions).

Third, 'old' technologies do not go quietly. In fact, they typically undergo repeated cycles of adjustment—improving performance and cutting costs—as they attempt to hold off rivals. The first steam powered vessels did not put an end to sailing ships. Instead, they ushered in a period of intense technical and commercial competition, which saw improved design and fabrication of sailing vessels, increased speed, and a reduction in crew size. In the end the decisive factor was not speed or cost, but the greater reliability of steam (Berkhout, Smith, and Stirling, 2003). Even when the age of sail had passed, wind-driven vessels remained in service in less developed regions. And eventually a new niche was carved out for sail in recreational, sports, and training applications. So 'old' technologies may survive long after their obituaries have been drafted. With respect to the energy sector the implication is that the oil and gas industries will not just roll up and die because some scientists claim to have come up with a nifty alternative fuel, or because others warn of long term environmental catastrophe. Oil and gas are too convenient; they remain too readily available; the technologies for their extraction and combustion are already highly perfected; and there is no reason to believe that options for further technological innovation (to increase extraction, increase efficiency, and mitigate environmental impacts) have been exhausted. Thus emergent renewable technologies will have to compete with a moving target: their rival is not just the oil and gas industry of today, but the fossil-fuel technologies of tomorrow.

Fourth, technological transitions have distributional consequences. The fortunes of individuals, companies, occupational groups, towns and geographic regions can be tied to the careers of specific socio-technological options. Change will always bring winners and losers. Even if society gains as a whole from the shift to a new technological trajectory, some groups will suffer. Jobs will be lost, and established skill sets may become redundant. The incomes, profits, or tax revenues, as well as the standing and influence of some groups will decline. Thus 'progress' has its costs, and its victims. And because such distributional consequences may be acute, technological transitions provoke resistance from established groups (companies, unions, professions, particular regions, and so on) that believe their interests to be threatened. From a policy perspective it is important to be aware of these realities. Broad coalitions may be required to isolate powerful groups that are opposed to change. And substantial social resources may have to be devoted to cushioning the impact of change both for powerful interests that might otherwise undermine the process, and for vulnerable groups that may be ill-equipped to bear the burdens of adjustment.

Fifth, while few technological transitions have originated as 'political projects', politics is closely intertwined with technological development at almost every level. Political circumstances influence what is possible, encouraging or discouraging investment and innovation both generally and in particular spheres. Policy intervention can protect technological incumbents or expose them to challenge. Interests associated with particular technological options consistently exploit the political realm to advance their projects. And governments, for their part, routinely lend support to technological initiatives which they deem to be in the public interest. Support for research and development in the military sphere is a given, but governments have also systematically backed the introduction of new infrastructure technologies that have captured the political imagination—such as railways, the electricity supply system, the highway network, and communications technologies. With respect to energy systems, the current landscape is influenced by policy at every level. So it is perfectly reasonable to ask whether the existing patterns of governmental intervention correspond to the public interest, and to contemplate reformulating policy to tip development in more socially desirable directions.

Of course, the transformation of energy systems will not represent a single 'technological transition', but a family of transitions in associated systems that will stretch out over time. This will involve not only the way energy is produced but also the way energy is consumed. After all, energy is not an end in itself, but a means to satisfy other needs—providing inputs for agriculture, construction, manufacturing, and transport, as well as direct services to households. Ultimately, changes to energy systems will have profound implications for the organization of industrial production, systems of transport, the spatial disposition of cities, and patterns of domestic life.

Critical policy considerations

So far the discussion has pointed to a number of elements that should inform efforts to govern energy transitions. Reference has been made to the idea of 'sustainable energy policy' as an appropriate policy frame. Emphasis has been placed on the broad range of societal factors involved in technological change. The uncertainty surrounding such processes, the advisability of focusing policy to attain functional goals (rather than privileging specific technological options), and the centrality of distributional conflicts have also been highlighted. Moreover, the point has been made that governments should not feel bashful about intervening in a domain where the configuration of existing socio-technical regimes has been so heavily influenced by previous rounds of policy choice, and where the potential impact on long term societal development is so high.

Keeping such considerations in mind, it is possible to go further in specifying the general parameters of the necessary approach. A critical element concerns societal engagement, dynamism, and inventiveness. On the one hand, citizens are *entitled* to have a say in decisions about defining the development trajectory 'beyond oil'. And, on the other hand, the successful negotiation of this change *requires* inputs from key stakeholders and the public. Since there are many possible energy futures—exploiting different technological options, implying different packages of economic, social and environmental costs and benefits, and involving different risks and opportunities—citizens should have some influence over the path that is ultimately chosen. Moreover, these are issues that can not be successfully 'sorted out' behind closed doors by scientists, industry leaders, and top state officials. Their reach is too profound; their impact on diverse social strata and practices is too great. And the innovation-potential required to address them is widely distributed. Thus the challenge is to progressively engage the public and key stakeholders; to enable them to appreciate the dilemmas, opportunities, and choices related to energy; and to mobilize their creative power to transform current practices in a more sustainable direction. The response to climate change and the transition towards a new energy system must be seen as defining challenges for our generation. And policy should be oriented to involve groups and individuals at all levels—schools, universities, professional associations, businesses, research labs, the media and artistic communities, charitable and religious groups, and so on. Energy and climate change should be set at the heart of the scientific and technological agenda, and the core of civic debate. They should be approached not in apocalyptic terms (we are running out of oil! the climate is going haywire!) but rather as defining elements of the sort of society we want for ourselves and for our children. Calls for enhanced 'public participation' in all sorts of policy contexts are fashionable at present, although officials are typically more concerned with formal than substantive processes. But in relation to energy and climate change what is needed is progress in general understanding (to provide a political underpinning for an active policy response) and in practical involvement (to provide multiple sources of innovation). Thus policy should be directed to engage and mobilize society, and particularly major stakeholders.

There is also a need for a clear orientation from central governments. This area cries out for strategic vision and long term planning. Not 'planning' in the sense of 'command and control dictates'—but 'planning' in the sense of forward-oriented analysis, that reviews trends, explores scenarios, and establishes priorities. Such a vision plays an important communicative function, both within government (letting officials at all levels know what is expected) and in relation to society at large (Meadowcroft, 1999). Precisely because processes of long term structural change are rife with uncertainty, societal actors (such as firms and households) need to understand the basic direction of government policy. For this allows them to orient their autonomous activities in relation to a long term perspective. A number of OECD states have recently made some progress in defining such visions, adopting longer term energy and climate change objectives. But to give such a strategic orientation life, it is necessary to integrate energy and climate change concerns into the work of departments across government, and to establish performance measures that hold organizations and officials accountable for their performance. Without such initiatives the governmental responses will remain fragmented and incoherent, with various ministries and agencies pulling in different directions. Of particular concern is the array of bodies with responsibility for energy regulation, especially in jurisdictions that have seen substantial deregulation of their energy markets. Sometimes the mandates of these institutions were defined almost exclusively in terms of competition policy, with no explicit reference to climate change and the long term energy transition. Another front on which a strategic orientation can guide action is with respect to the comprehensive review of existing energy-related policies and expenditure. Much of existing energy policy reflects the accretions of an earlier age, and it includes subsidies for fossil fuel industries and carbon-intensive sectors, and encouragement of energy wasting practices we now recognize as perverse. Cutting established subsidies and tax concessions is always a political challenge, but the attempt must be made to re-orient public expenditure in line with current priorities.

Another key consideration is that, in making energy policy, as much attention should be paid to usage and demand as to production and supply. So long as energy systems generate significant environmental costs, it is better to use less energy than more. In particular, while the energy mix continues to include sources that release greenhouse gases to the atmosphere, reducing demand helps control emissions. This implies an active effort to promote energy efficiency, and to manage rebound effects in order to secure demand reductions. This is not so much about behavioural change in a static technological context (lowering speed limits, extending daylight savings time, or urging householders to turn down the thermostat and put on extra sweaters)—as it is about encouraging socio-technical innovation on the consumption side. This means driving the shift to more efficient heating, cooling, and lighting in domestic and commercial contexts. It means encouraging industrial innovation to reduce the energy intensity in manufacturing and construction. Possibilities for significant demand reduction in the medium term are good, with innovation in the design of buildings, improved energy consumption of appliances and electronic equipment, and the transformation of industrial processes and materials towards lower energy pathways. But operating on the demand side is more difficult than concentrating on new supply technologies, because consumption is fragmented among many types of user (industrial, commercial, domestic). Moreover, technologies of 'saving' may appear less glamorous than technologies of production. Yet if energy savings are linked to cost efficiencies, and to gains in functionality, they can appeal. And policy oriented to encourage movement in this direction is urgently required.

Also important is the design of packages of policy instruments suited to different dimensions of the problem. Over the past three decades there has been an enormous accumulation of knowledge about the deployment of different instruments in the fields of environment and energy (e.g. Golub, 1998; Harrington, Morgenstern, and Sterner, 2004). To the established understanding of regulatory governance has been added experience with market-based instruments, negotiated approaches, and informational techniques (OECD, 2003). The use of mechanisms such as portfolio standards, renewables levies, feed in tariffs, tradable emissions permits, accelerated depreciation allowances, and product labelling, have shown the potential and the limitations of specific approaches, and the conditions where they are more or less likely to succeed. A key challenge is combining different sorts of instruments into balanced packages to achieve specific goals. With respect to science and innovation policy one important lesson has been that different instruments are required at distinct phases of the research and innovation cycle. Some measures can stimulate primary research, while others are more appropriate to encouraging applications-development, product roll-out, or consumer up-take. Attention must be devoted to identifying obstacles in innovation pathways, including obstacles on the business side (operation of venture capital markets, business models and practices) and on the government side ('red-tape', procurement policies, outdated standards, jurisdictional tangles).

Exercises combining the visioning of alternatives, practical experiments and networking—that assemble new constellations of actors interested in emergent technologies—are particularly important. They can help identify obstacles to, and opportunities for, innovation that cannot be appreciated by groups working in isolation—be they from industry, government, or civil society. Such initiatives are central to the approach that has gained ground in the Netherlands over the past few years under the label of 'transition management' (Rotmans, Kemp, and van Asselt, 2001). Formulated initially by researchers working on the Fourth National Environmental Policy Plan, the idea has been taken up to orient the long term movement towards sustainability in key sectors—energy, transport, natural resources, and so on (NEPP4, 2002). The Dutch Ministry of Economic Affairs has been particularly active with respect to the energy transition, funding an array of projects that explore socio-technological alternatives to current patterns of production and consumption. The preparation of a report presenting energy scenarios through to 2050 provided the basis for identifying five strategic transition 'routes' (green and efficient gas, enhanced production chain efficiency, green raw materials, alternative motor fuels, and sustainable electricity) that were considered 'robust' across varied scenarios (Bruggink, 2005). Further consultation with stakeholders allowed the formulation of aspirational goals ('ambitions'), transition paths (strategies for change), and specific 'options' (technological and social innovations) for each strategic route. Projects organized by coalitions of stakeholders have been funded to explore transition paths (Kemp and Loorbach, 2005). For example, one project focused on reducing energy usage in the paper and board sector by 50% by 2020, while another engaged with the agricultural glasshouse industry. An evaluation of existing research funding from the perspective of transition management has been undertaken, and the Dutch government has established a 'Frontrunners desk' to cut through 'red tape', reduce the regulatory burden on proactive firms, and identify bureaucratic obstacles to novel experiments.

'Transition management' has been described as 'a deliberate attempt to bring about long term change in a stepwise manner, using visions and adaptive, time limited policies' (Kemp and Rotmans, 2003). Interactions among concerned stakeholders are central to the iterative processes at the heart of transition management. Concerned parties are drawn into continuing discussions about goals and visions, the identification of interim objectives, and the assessment of progress. Thus transition management appears as a further extension of the interactive modes of environmental governance already institutionalized in the Netherlands.

Although particulars of this approach might be hard to reproduce in a country that does not have such a 'consensus-oriented' and 'planning-friendly' political culture, the fundamental impulse of 'transition management' is of general significance. For the establishment of networks involving novel constellations of public and private groups focused on innovation in the energy domain is vital. And these can only be truly effective if they operate within a context where government provides a clear strategic orientation signalling its commitment to long term change in the energy sphere.

Transforming energy/environmental governance

When discussions of energy futures get underway there is typically a tension between two opposed perspectives. On the one side there are the enthusiastic advocates of alternative technologies—scientists, engineers, environmentalists and, entrepreneurs—convinced they have viable options that can meet society's energy needs. From them one gets the impression that if only a few obstacles (technological, economic, or political) could be overcome, these new approaches would flourish. On the other side, there are the old energy hands, experts in existing technologies and markets, who emphasize the overwhelming place fossil fuels occupy in our present energy mix, and who disparage the possibility of any early or rapid shift towards alternatives. To the first group, the arguments of the second appear backward (seeming to ignore that that the days of oil are numbered), or worse, as an apology for the powerful interests that currently dominate the world energy order. To the second group, the arguments of the first are those of naïve dreamers, special interest pleaders, or 'jonnie-come-lately' analysts who don't understand the real economics of energy, or the mountains alternative technologies will have to climb to present more than a trivial challenge to the overwhelming predominance of fossil fuels.

In fact, there is truth on each side of this divide. It is easy to become wrapped up with the potential of new technologies without appreciating the obstacles to their widespread adoption, the costs of transition, the problems a shift might eventually entail, and the long lead times between proof of concept and societal deployment.

Moreover, since choices are made relative to other options, the emerging energy technologies described in this volume will often be as much in competition with each other as they are with the core of the established fossil fuel industry. Clearly change will take time, and some promising alternatives will not pan out. And yet it also easy to be mesmerized with the current state of the world, and to be overawed by the solidity of established socio-technical systems. Just because one practice holds sway today does not mean that it will always be ascendant. Indeed, as we have seen in the energy domain, the factors driving change over the medium term are destined to grow. Within two or three decades the impact of new renewables, cumulative efficiency and demand control innovations, and carbon storage technologies could be enormous. Although this represents quite a long period in the life of an individual, it is short in relation to the duration of the fossil era or to the time scale over which climate management efforts will be required. And it is well within the potential planning horizons of major corporate and governmental actors.

Yet if governments are to steer societies in such a direction, they will have to move beyond the approach to energy and environmental issues that characterized the final decades of the twentieth century. A number of necessary changes have been discussed over the course of this chapter. But two require further elucidation. First, there is an urgent need to integrate different dimensions of decision making. Energy policy, environmental policy, and science and innovation policy must be brought into closer contact. Energy solutions can no longer be defined without taking proper account of environmental factors (especially those related to climate change), while the advance of science and technology are critical to understanding and problem-solving in both the environmental and energy domains. Moreover, there are essential links between each of these three areas and sectoral policies—focused on transport, construction, agriculture, industry, and so on. The idea of 'environmental policy integration', which has been much discussed over recent years and formally adopted as a goal in many jurisdictions (such as the European Union), captures part of this dynamic. So too does the more encompassing notion of sustainable development, which insists that economic, social, and environmental elements be brought together in decision making. But achieving such 'integration' in practice is more difficult than achieving it rhetorically. And as the earlier discussion has suggested, what is required is not a 'defensive' integration—that simply subjects distinct sectoral policies to environmental review, and attempts to define supplementary measures to mitigate impacts. Rather environmental policy should contribute to redefining the desired sectoral development trajectories, while innovation policy accelerates technological advance in critical areas.

Second, there is the move towards more 'interactive' modes of governance. As the recent literature in political science have documented, the contexts in which governments operate have changed markedly over the past three decades.

The situation is more complex and turbulent, the range of actors and the scope of issues have grown, and economic and political ties among states are more intimate (Pierre, 2000; Pierre and Peters, 2000). Governments now place more emphasis on interactive approaches, relying on the 'self governing' potential of social organizations and 'co-governance' arrangements (Kooiman, 2003). Nowhere is this more important than in the environmental and energy domains. Governments cannot 'solve' energy problems on their own. But neither will these problems simply disappear if they are left to market mechanisms, or voluntary initiatives. The state remains a powerful instrument for achieving collectively defined ends, but the way it defines and realizes those ends is changing (Eckersley, 2004; Barry and Eckersley, 2005). This does not mean that regulation and expenditure—the two staples of governmental action—are no longer required. But it does mean that it is at the interface among different types of organizations— including governmental agencies, corporations and business associations, and civil society groups—that much of the more dynamic work of governance goes on (Driessen and Glasbergen, 2002; Meadowcroft, 2004). In particular, these interactions can play an important role in spurring the social and technological innovations required to manage energy transitions.

Throughout this chapter it has been argued that the technological developments required to meet the energy and environmental challenges of coming decades are as much about societal innovation and transformation as they are about scientific and technological discovery. And political processes and policy decisions play a key role in shaping the way societies create and use energy. For two centuries economic advance has been closely tied to fossil fuel combustion. Not only do hydrocarbons meet the bulk of our energy needs, they also provide the feedstock for the modern materials economy—plastics, chemicals, fertilisers, and so on (Geiser, 2001). Our civilization has prospered by tapping the solar energy captured by living organisms and laid down over geological time scales. But now worries about the supply of the most convenient fossil resources and above all a growing appreciation of the environmental consequences of current practices suggest that energy systems must change. Over coming years governments and citizens will face critical choices about energy options and about how serious they are about tackling the issue of climate change. Sooner or later we must move 'beyond oil'. Clearly there are many possible energy futures— with some far less desirable than others. And by making wise political choices today we can help influence which of these alternative futures actually becomes reality.

References

Barry, J. and Eckersley, R. (eds.) 2005. *The State and the Global Ecological Crisis*, MIT Press, Cambridge, Mass.

Berkhout, F. and Gouldson, A. 2003, Inducing, shaping, modulating: perspectives on technology and environmental policy., In F. Berkhout, M. Leach and I. Scoones (eds.) *Negotiating Environmental Change: New Perspectives from Social Science*, Edward Elgar, Cheltenham..

Berkhout, F., Smith, A., and Stirling, A. 2003. Socio-technological regimes and transition contexts, *SPRU Electronic Working Paper Series 106*, SPRU, Brighton, available at: http://www.sussex.ac.uk/Units/spru/publications/imprint/sewps/sewp106/xsewp 106.pdf.

Bijker, W. (1995), *Of Bicycles, Bakelites, and Bulbs: Toward a Theory of Sociotechnical Change*. MIT Press, Cambridge, MA.

Bijker, W., Hughes, T., and Pinch, T. (eds.) 1989. *The Social Construction of Technological Systems: New Directions in the Sociology and History of Technology*. The MIT Press, Cambridge, MA.

Boyle, G. (ed.) 2003., *Renewable Energy: Power for a Sustainable Future*, Oxford University Press and the Open University, Oxford.

Boyle, G., Everett, B., and Ramage, J. 2003. *Energy Systems and Sustainability: Power for a Sustainable Future*, Oxford University Press and the Open University, Oxford.

Bruggink, J. 2005. *The Next 50 years: Four European Energy Futures,* Petten: Energy Research Centre of the Netherlands.

Doern, B. (ed.) 2005. *Energy Policy and the Struggle for Sustainable Development*, University of Toronto Press, Toronto.

Doern, B. and Gattinger, M. 2004. *Power Switch: Energy Regulatory Governance in the Twenty-First Century*, University of Toronto Press, Toronto.

Driessen, P. and Glasbergen, P. 2002. *Greening Society: the Paradigm Shift in Dutch Environmental Politics*, Kluwer, Dordrecht.

Eckersley, R. 2004. *The Green State: Rethinking Democracy, and Sovereignty*, The MIT Press, Cambridge, Mass.

EEA (European Environment Agency) 2002, *Energy and the Environment in the EU*, EEA, Copenhagen.

Geels, F. 2002. *Understanding the Dynamics of Technological Transitions: A Co-evolutionary and Socio-technical Analysis*, Twente University Press, Twente.

Geiser, K. 2001. *Materials Matter: Toward a Sustainable Materials Policy*, MIT Press, Cambridge, Mass.

Golub, J. 1998. *New Instruments for Environmental Policy in the EU*, Routledge, London.

Harrington, W., Morgenstern, R., and Sterner, T. 2004. *Choosing Environmental Policy: Comparing Instruments and Outcomes in the United States and Europe*, Resources for the Future.

Harris, M. 2002. *Energy Market Restructuring and the Environment: Governance and Public Goods in Globally Integrated Markets*, University Press of America, Lanham, Md.

IEA (International Energy Agency) 2004. *World Energy Outlook 2004*, OECD, Paris.

IMF (International Monetary Fund) 2005. *World Economic Outlook*, IMF, Washington.

Kemp, R. and Loorbach, D. 2005. Dutch Policies to Manage the transition to Sustainable Energy. In *Jahrbuch Ökologische Ökonomik 4 Innova-tionen und Nachhaltigkeit*, MetropolisVerlag.

Kemp, R. and Rotmans, J. 2003. Managing the transition to sustainable mobility. in B. Elzen, F. Geels and K. Green (eds.), *System Innovation and the Transition to Sustainability: Theory, Evidence and Policy*, Edward Elgar, Cheltenham.

Kooiman, J. 2003. *Governing as Governance*, Sage, London.

Lafferty, W. (1996) 'The politics of sustainable development: global norms for national implementation', Environmental Politics 5: 185–208.

MacKenzie, D. and Wajcman, J. (eds.) 1999. *The Social Shaping of Technology* Open University Press, Buckingham.

Meadowcroft, J. 1997. Planning for sustainable development: insights from the literatures of political science., *European Journal of Political Research* **31**: 427–54.

Meadowcroft, J. 1999. Planning for sustainable development: what can be learned from the critics? In M. Kenny and J. Meadowcroft (eds.) *Planning for Sustainability,* pp. 12–38. Routledge, London.

Meadowcroft, J. 2004. Participation and sustainable development: modes of citizen, community, and organizational involvement. In William Lafferty (ed), *Governance for Sustainable Development: The Challenge of Adapting Form to Function*, pp. 162–90. Edward Elgar, Cheltenham.

NEEP4 (Forth National Environmental Policy Plan of the Netherlands) 2002. *Where There's a Will There's a World*, Ministry of Housing Spatial Planning and the Environment, the Hague.

OECD 2003., *Voluntary Approaches for Environmental Policy: Effectiveness, Efficiency and Usage in Policy Mixes*, OECD, Paris.

Pierre, J. (ed.) 2000. *Debating Governance: Authority, Steering and Democracy*, Oxford University Press, Oxford.

Pierre, J. and Peters G. 2000 *Governance, Politics and the State,*: Macmillan Press, London.

Plourde, A.2005. The changing nature of canadian and continental energy markets. In 'IPCC, Climate Change 2001: Impacts, Adaptation, and Vulnerability,' *Summary for Policymakers, A Report of Working Group 11*.

Ristinen, R. and Kraushaar, J. 1999. *Energy and the Environment*, John Wiley and Sons, New York.

Rotmans, J.,Kemp, R., and van Asselt, M. 2001. More evolution than revolution: transition management in public policy, *Foresight, 3*, 15–31.

Sustainable Industries Journal. 2005. 'Sweden declares war on oil', November 2005, available at: http://www.sijournal.com/energy/1914117.html.

WCED (World Commission on Environment and Development) 1987. *Our Common Future*, Oxford University Press, Oxford.

The author

James Meadowcroft holds a Canada Research Chair in Governance for Sustainable Development and is Professor in the School of Public Policy and Administration and in the Department of Political Science at Carleton University. His research focuses on reforms to structures and processes of governance to promote the transition towards sustainability. After completing a BA in Political Science at McGill, he obtained a doctorate from Oxford University. Recent contributions include work on public participation, sustainable development partnerships, planning for sustainability, national sustainable development strategies, and sustainable energy policy.

14. *Summary*

Robert May

Energy...Beyond Oil is important and timely and should be understood within the wider context of global climate change and future energy demands.

In the 1780s John Watts developed his steam engine and so began the Industrial Revolution. At this time, ice-core records show that levels of CO_2 in the atmosphere were around 288 parts per million (ppm). Give or take 10 ppm, this had been their level for the past 6,000 years, since the dawn of the first cities. As industrialization drove up the burning of fossil fuels in the developed world, CO_2 levels rose.

At first the rise was slow. It took about a century and a half to reach 315 ppm. The rise accelerated during the twentieth century: 330 ppm by the mid-1970s; 360 ppm by the 1990s; 380 ppm today. This change of 20 ppm over the past decade is equal to that last seen when the most recent ice age ended, ushering in the dawn of the Holocene epoch, 10,000 years ago. If current trends continue, then by about 2050 atmospheric CO_2 levels will have reached around 500 ppm, nearly double pre-industrial levels. The last time our planet experienced such high levels was some 50 million years ago, during the Eocene epoch, when sea-levels were around 100 m higher than today. The Dutch Nobelist, Paul Crutzen, has, indeed, suggested that we should recognize that we are now living in a new geological epoch, the Anthropocene. He sees this epoch as beginning around 1780, when industrialization began to change the geochemical history of our planet.

Even today, there continues to exist a 'denial lobby', funded to the tune of tens of millions of dollars by sectors of the petrochemical industry, and highly influential in some countries. This lobby has understandable similarities, in tactics and attitudes, to the tobacco lobby that continues to deny smoking causes lung cancer, or the curious lobby denying that HIV causes AIDS. This denial lobby is currently very influential in the USA. It is astonishing that a country that can squander roughly one hundred billion dollars on the Star Wars 'missile-defence' programme (which has not produced a working system, and would probably diminish global security if it did!) should shy away from acknowledging the dangers of climate change. No other species in the history of life on earth has faced a problem of it own creation that is as serious as we now do. Earlier, when some aspects of the science were less well understood, they denied the existence of evidence that human inputs of CO_2 and other greenhouse gases were causing global warming. More recently, there is acknowledgement of anthropogenic climate change, albeit often expressed evasively, but accompanied by arguments that the effects are relatively insignificant thus far and/or that we should wait and see, and/or that technology will fix it anyway.

It is hard to avoid the conclusion that climate change is real and caused by human activities. This has been affirmed by the Inter-Governmental Panel on Climate Change (IPCC[1]), which comprises the world's leading scientists, by the US National Academy of Sciences (in its 2001 report), and most recently by a statement[2] from all the Academies of Science of all G8 countries, along with China, India, and Brazil. This statement, deliberately designed to clarify the consensus on climate change for the Summit Meeting under the UK Presidency of the G8 in July 2005, calls on the G8 nations to: 'Identify cost-effective steps that can be taken now to contribute to substantial and long-term reduction in net global greenhouse gas emissions. [And to] recognise that delayed action will increase the risk of adverse environmental effects and will likely incur a greater cost'. In short, it is clear that world leaders, including the G8, can no longer use uncertainty about aspects of climate change as an excuse for not taking urgent action to cut emissions of CO_2 (and other greenhouse gases).

The impacts of global warming are many and serious: sea level rise (which, owing to long lags in the warming of deep water, would continue for centuries even if CO_2 emissions returned to pre-industrial levels tomorrow); changes in water availability (in a world where human numbers already press hard on available supplies in many countries); and the increasing incidence of 'extreme events'— floods and droughts—the serious consequences of which are rising to levels which invite comparison with 'weapons of mass destruction'. As emphasized at a recent Royal Society meeting on climate change and crop production, 'Africa is consistently predicted to be among the worst hit areas across a range of future climate change scenarios'. This particular fact underlines a grim resonance between the UK Government's two themes for its G8 Presidency: climate change and Africa.

[1] http://www.ipcc.ch
[2] http://www.royalsoc.ac.uk/displaypagedoc.asp?id = 20742

So, what should we be doing? One thing is very clear. The magnitude of the problem we face is such that there is no single answer, but rather a wide range of actions must be pursued. Broadly, I think these can be divided into four categories. First, we can adapt to change: stop building on floodplains; start thinking more deliberately about coastal defences and flood protection, recognizing that some areas should, in effect, be given up—in Holland, one quarter of which lies below sea level, there are already plans for houses designed to float on seasonally flooded areas. Second, we can reduce inputs of CO_2 by reducing wasteful energy consumption. Chapter 12 of this book, entitled *Energy Efficiency in the Design of Buildings* by Dell and Egger, notes that we could design housing which consumes less than half current energy levels without significantly reducing living standards. Wasteful consumption of energy abounds in the developed world, at home, at work, and in the marketplace. Third, we could capture some of the CO_2 emitted in burning fossil fuels, at the source, and sequester it (burying it on land or under the sea bed). Fourth, we could embrace the many different renewable sources of energy described in preceding chapters of this book, which do not put greenhouse gases into the atmosphere.

For most people, the only really satisfactory way to generate energy is by magic! Because every practical way has offensive elements to it, and much of the discussion then is about the relative offensiveness, because everything has a problem unless it's a yet unproved technology, which if you're 'wishy' enough and self-indulgent enough, you can look to as the answer. But of course there is no magic solution and there is no single, simple technological fix.

Following the first report of the IPCC in 1990, the Earth Summit in Rio de Janeiro in 1992 addressed the issue of climate change. The consequent Framework Convention of Climate Change was agreed by more than 160 countries, and signed by the first President Bush for the USA. It stated that the Parties to the Convention should take: 'precautionary measures to anticipate, prevent or minimize the causes of climate change and mitigate its adverse effects. Where there are threats of irreversible damage, lack of full scientific certainty should not be used as a reason for postponing such measures'. All nations need to take part in such reductions in emissions of CO_2. There are, however, very large differences among the levels of emission in different countries. In terms of metric tons of carbon input to the atmosphere per capita per annum, the variation is from about 5.5 for the USA, 2.2 for Europe, 0.7 China, 0.2 for India, and down to even lower levels for many developing countries. For the past several decades, the developed world has been moving—to different degrees in different countries—from coal to oil, gas and (to a small extent) renewables. The resulting lower CO_2 emissions per energy unit is known as 'decarbonization'. But with the rapid growth in energy usage that is set to continue in industrializing countries like China and India, where supplies of coal are far more abundant than oil or gas, the next few decades are likely to see unhelpful 're-carbonization'. Indeed, China is expected to surpass the USA as the world's largest carbon emitter by around 2025.

When we were hunter-gatherers, up to about 10,000 years ago, our consumption of energy was what we could catch and what the women, who did most of the work then, as now, could grow, and we just got enough for basic metabolic processes. In contrast today, the average inhabitant of the globe—and there's a great deal of variance among different places—consumes in daily life 15 times the amount of energy needed to maintain basic metabolic processes in daily life, and even in the most impoverished areas, it's a factor of several compared to our earlier ancestry—which had much in common with the rest of the living world.

We've increased in numbers, which is all part of this problem. As hunter-gatherers, there were probably only about 10 to 20 million humans—we couldn't have supported more. It was the dawn of agriculture that set us on a path which, with bumps and curves in it, took us to the first billion in 1830. As the Industrial Revolution gathered steam, we doubled in 100 years, and we doubled again in 40 years, 4 billion in the early 70s, and we're 6.4 billion now. Conservative estimates suggest another 50%, all of them added in big cities, by the middle of the current century.

An estimate of the footprint humanity casts on the globe, the consumption of energy and other of the materials in life, varies from country to country. The country whose inhabitants have the biggest average footprint is the United Arab Emirates, with the USA a close second. Consumption in those countries is something like six times the average inhabitant of China, twice that of Europe. If you look at that aggregate footprint, the average consumption patterns in different regions, multiplied by the number of people, and compare it to what is the best estimate of what the Earth can sustain, you find we have just recently crossed this threshold. So to put it another way, there's no point in discussing what if China were to move to our consumption patterns, because it is not going to happen. Whatever the imprecisions in that estimate, if China were to reach US consumption patterns, it would exceed by a clear factor of 3 or more what the planet can sustainably support in the relatively short term.

To return to *Energy . . . Beyond Oil,* there are two 'problems' implicit in our title: for one thing, production of oil is expected to pass its peak, and decline, perhaps in the next few years, maybe not for several decades, but 'soon' on any reasonable time-scale; for another thing, oil accounts for 35% of global energy production (in 2001), and hence is a major factor in climate change. Some people regard these issues as Bad Things. While I firmly agree that both problems are important, I think the first—oil supplies declining—may actually be a Good Thing. However great the difficulties it causes, it is the kind of seismic shift that will really drive home, in ways that cannot be disregarded, that we live on a finite planet and that we need to think more deliberately about our future.

As the reader will see, this book surveys the full spectrum of possible near term and long term sources of renewable energy generation. Some of these are already technically advanced, others are as yet in the preliminary stage of development, and some may even appear as visions for the future. Many have their own serious problems, albeit different from those causing planetary warming. Some, particularly nuclear fission (7% of global energy production in 2001), currently play an important part as non-CO_2-producing energy sources. Others, especially nuclear fusion, hold the promise as commercially viable sources of power in a more distant future. Vagaries of fashion seem to smile on some and frown on others. It is not clear to me why such determined effort is put into wind turbines, whilst so little is put into developing the technology of geothermal energy and wave energy sources.

Renewable and emissions-free energy sources are only one part—albeit an important part—of a necessarily multifaceted approach to addressing climate change. To get some idea of the ultimate magnitude of the challenge, note that in 2001 90% of all energy generation emitted CO_2 into the atmosphere (79% from fossil fuels, 11% from biomass), with only 10% (7% nuclear fission, 3% renewables) from non-CO_2-emitting sources.

But there remains a big gap for the 80 to 90% of energy generation that fossil fuels represent at the moment: 7% of global energy, two-thirds of the non-putting CO_2 into the atmosphere, and 17% of the electricity comes from fission. In the UK, we've declined in recent years from 29%, down to 22%. In Germany, they have renounced it, and when asked, 'Well, what will make up?' they buy in the fission-produced electricity from France.

I have every sympathy with those who have a fundamental objection to nuclear power. It is understandable to see it through the emotional haze of a mushroom cloud. And there are real problems with terrorism and waste disposal that add to that. I have a lot of sympathy with these worries, but I have difficulty seeing, much as I wish it were otherwise, how we're going to get through the medium term without nuclear energy. I believe, and I think it is not simply wish fulfilment, that the estimate of having proof of concept for nuclear fusion by 2031 may be realistic, and then 5 to 15 years on, routine energy production, and that would really be helpful and should be striven for. Let's hope we keep it going with the momentum it's got at the moment.

It is not a syllogism to say: 'I am an environmentalist and therefore I'm against nuclear energy'. If you're an environmentalist, therefore you're for the environment and you are opposed to the threat that climate change poses. So ask what can be done, recognizing every choice has its problems. We can differ about the details, but don't assume a dogmatic position about any one of the energy solutions.

It is imperative to recognize that there is no single answer to the problem of *Energy . . . Beyond Oil*. Each and every one of the solutions described in the chapters of this book is a potential part of the future.

The author

Robert M. May (Lord May of Oxford) OM AC FRS, has been Chief Scientific Adviser to the UK Government (1995–2000) and President of the Royal Society (2000–2005). His research interests include mathematical ecology, chaos theory, biodiversity, and extinction. He has been awarded numerous international awards, medals, and honorary degrees, including the Royal Swedish Academy of Science's Crafoord Prize, the Balzan Prize, and the Blue Planet Prize.

Index